Genetic Transparency?

Life Sciences, Ethics and Democracy

Edited by

Peter Derkx
Erica Haimes
Harry Kunneman

VOLUME 2

The titles published in this series are listed at *brill.com/lsed*

Genetic Transparency?

Ethical and Social Implications of Next Generation Human Genomics and Genetic Medicine

Edited by

Malte Dreyer
Jeanette Erdmann
Christoph Rehmann-Sutter

BRILL
RODOPI

LEIDEN | BOSTON

Cover illustration: Tobias Haberland.

Library of Congress Control Number: 2015958270

ISSN 2211-4416
ISBN 978-90-04-30668-4 (hardback)
ISBN 978-90-04-31189-3 (e-book)

This book is printed on acid-free paper and produced in a sustainable manner.

Contents

Introduction

Christoph Rehmann-Sutter, Malte Dreyer, Jeanette Erdmann

It is historically unprecedented for people to be able to know personal genomic information. Not just information about oneself, retrieved from the nuclei of one's body cells through sequencing or testing, but also information about other individuals – children, unborn children, relatives, and other people. Who should have access? To whom should genetic information be available and in what form? Answers have been given, and the laws that have been created over recent years, use the language of rights and duties. People cannot be denied the right to know what other people could know about them. And if there is a right (rather than a duty) to know about one's genome, there must be also a right not to know. But personal (moral) duties to know might nonetheless emerge, even within the framework of a legally granted right not to know. This changes the way we live individually, the way we live together, the shape of society. Part of this revolutionary force of genomic knowledge may even arise inadvertently, as consequences of choices that are made on other grounds, for different reasons and with other aims. Human life can be changed simply as a consequence of the possibility of knowing, just because genetic information is within reach. This situation leads to a number of questions about genetic knowledge:

Who is entitled to have genetic knowledge? Who is claiming or appropriating knowledge about whom – and about what? How are data interpreted, transformed, read – or misread? And how do people make sense of their data in everyday life? How is the lack of knowledge sometimes also meaningful – in a situation where one could know?

Novelist and journalist Carole Cadwalladr wrote, "It's hard to overstate how new genomic science is, how quickly it's changing, how revolutionary it's likely to be,"[1] after she had her whole genome sequenced and was still excited at knowing something about herself that most people cannot know, but was also becoming aware of the limits and ambivalence of this unprecedented knowledge. The emphasis she gave is twofold. First, that genomic sciences are novel, rapidly changing and revolutionary for societies and for their healthcare systems; and second, that this is equally true on a personal level for individuals and families, people like Cadwalladr herself who have obtained some insight into their genome. What is it exactly that genetic knowledge is about? In the same article Cadwalladr gave one answer: its insight into the "complete code of everything in our bodies".[2] Perhaps this expectation is too high and it reflects more the culturally motivated gene-centrism that became somewhat hegemonic in the late 20th century. Even if we disagree with the definition of the genome as the "complete code of everything in our bodies" and lean towards a more reduced, modest, even a sceptical view, we see that objective genetic information always has a subjective side. We say that genomic knowledge becomes subjectified, even if it is only anticipated knowledge. The subjective interpretation is an act of meaning-generation that matters, an activity that can be reflected upon and critically discussed. This is beyond science. Genomes are (and always have been) interpreted culturally and philosophically, even theologically, as something with a meaning for us.

The accessibility of genomic information creates a kind of transparency of the human body to ourselves and possibly also to others. Speaking of 'transparency' in this context is of course metaphorical. Genomics does not make the body transparent in the literal sense. But genetic information changes the way we see our bodies and our embodied life. A genetic gaze will be practised increasingly culturally; knowing concretely which series of mutation, with an associated disease risk, we carry changes the way we see our present and future life. Hence, the visual metaphor is not arbitrary. It is however important to keep in mind that what we call here 'genetic transparency' is not only revelatory and informative because it brings

---

[1] Cadwalladr, 2013.

[2] Ibid.

hidden aspects into light, but also transformative in what we can 'see' and how we 'see'.

There is an ambiguity in the term 'transparency' which was useful for our writing project. In certain contexts, transparency is a positive and desirable feature. In other contexts, however, transparency is a threat to privacy, a danger to what we find desirable. In both regards, transparency is a political term that can be applied to bureaucracies, management structures, decision making, power. The ambiguity of the term is connected to its etymology. The word 'transparency' is composed of two Latin words trans (= through) and parere (= to appear or to become visible). It is therefore not surprising that we associate transparency not only with trust, but also with surveillance and control.

Political decision-making can be transparent (or lack transparency) for citizens who are affected by it; the procedures of the management of a multinational company can be opaque to those at ground level (staff, customers) who are affected by them. Citizens demand transparency about public finances because it is their taxpayer money. This is one side of political transparency: transparency as a criterion of good private and public management. Open government is an umbrella slogan for the principles that public institutions ought to use. When public trust is crumbling, politicians often declare they will build trust through transparency. In doing so they tend to ignore the fact that trust means building a positive relationship with somebody, which can endure in spite of not knowing everything about this person's actions. "It enables us to act with others, despite the lack of knowledge about details. [...] Transparency means visibility and the elimination of ignorance. More transparency means less room for trust."[3] Hence, the call for transparency can be a symptom of a broken or a damaged relationship, and at the same time the prolongation of mistrust.

In power relationships we have roughly two parties: those exercising power and those affected by power. What we have said above was about transparency from the perspective of those affected by those exercising power. Transparent governance was the issue. What about the tendency towards making citizens' behaviour

---

[3] Schües, 2013, 71 (our translation).

transparent to those exercising power over them? When citizens' private spheres are disregarded, say in order to make them more susceptible to preventive healthcare governance (i.e. supposedly in their best interests), the legitimacy of a breach of private spheres rests on the consent of these citizens. Consent both provides and limits this legitimacy. Citizens cannot legitimately be made genetically transparent in order to improve their health without their consent. Why is health improvement not a sufficient reason to justify transparency? The reason is that we have different ideas and ideals about health in our lives. Philosophers try to capture this with the idea of the "good life". Ideas of what is desirable and meaningful in life differ in a society, and healthcare governance needs to respect these differences, otherwise healthcare would become paternalistic and coercive.

Another contextual factor needs to be taken into account. This book discusses issues related to genetic transparency in a time marked by a series of scandals about covert digital surveillance by powerful agencies such as the US National Security Agency (NSA), or the UK Government Communications Headquarters (GCHQ), which were brought to light by whistleblower Edward Snowden and the Wikileaks website. The shock of these revelations was not only that surveillance at this level exists, but also that the public was not informed, not even in principle, about those activities. This links these issues with many other instances of surveillance through access to personal data that we use in our daily lives. The algorithms that control money and information today are inaccessible to most of us. Secrecy goes so far that most of us are completely unaware of the existence of such powerful algorithms. Transparency about the power structures is however a political demand that relates to justice and human rights. This applies to the algorithms that control data access.[4]

At the same time, genetic technologies enable more and more people to access their own personal genomic information. The genome can be sequenced – on demand. There are so-called direct-to-consumer gene test companies that offer tests on the basis of a simple saliva sample. The genetic data that are generated by such firms and communicated to their customers are stored somewhere

---

[4] Pasquale, 2015.

and used somehow. Can we trust them? If privacy is dead – as many today seem to believe – what should we think about genetic privacy? Can 'Big Data' in biomedical research empower the public – not just researchers – without endangering privacy? Are there new ways of building trust in healthcare providers and to science in a situation where scientists can no longer guarantee patients' privacy?[5] Genetic information and genetic knowledge, together with other big data-driven research that makes it possible "to connect the dots" from different registries and to generate individual, personalized profiles from a diversity of sources, is a big challenge for societies, institutions, families and individuals. In 2013, a proof-of-principle study drew public attention. It demonstrated that it is possible to re-identify the individual even from anonymised genome data. Yanik Erlich and his group used only free, publicly available internet information to find out the identity of the donor of the genome that had been anonymously sequenced in a research study.[6] This possibility has consequences for the way participants in genome studies need to be informed, even if they, or at least some of them, may not see re-identification of sequence data with their name as their primary concern.[7] There is no such thing as an "anonymous genome". The genome can always, at least in principle, be re-identified. It is like strewing our body cells around in the knowledge that everybody, in principle, could sequence the genome and find out who it was who lost these cells. This may not be damaging. But we can easily imagine circumstances in which we would strongly prefer our genomic information (or a particular detail of it, such as proving or disproving paternity, a mutation that represents a genetic disease risk etc.) not to fall into the wrong hands.

This applies beyond healthcare. We are afraid of becoming transparent to the eyes of all-powerful parties who may have an interest in us, in taking advantage of us in some way. Transparency in this regard is negative. It is a threat and a risk. How do we react in these circumstances? If information can fall into the wrong hands, we

---

[5] Wible/Jasny, 2015.

[6] Gymrek et al., 2013; Hayden, 2013.

[7] This is the result of a recent study with female patients in personalised breast cancer therapie. See: Rogith et al., 2014.

usually ask for rules to govern potential disclosure, and for good reason. And we want both to know about those rules and have a say in them. We approve of some of the hidden eyes who could look at our private information, but not of others. In the first place, therefore, we need to investigate what kinds of questions of genetic transparency must be discussed in a democratic society.

David Brin, an author who has analysed what transparency means in our information era, made a basic suggestion about issues of informational freedom: when it is under threat the best answer for individuals is not secrecy. This would have harmful consequences because secrecy would lead to a society of mistrust and arrogant elites – a society in which we would not really want to live. Instead, we should participate in *negotiating the rules that govern transparency*. A transparent society is at least transparent with regard to transparency, which presupposes the accountability of those who could acquire power through accessing data. Accountability, for Brin, means five things: that those who could misuse information must be known; when they misuse information this should be disclosed; the rules governing their use of information must be negotiated beforehand; power over the use of information must be distributed; and participation in the negotiations must be possible. We have suggested elsewhere that in principle these suggestions also apply to genetic information.[8] But many details of this proposal will need to be clarified and discussed.

A diversity of perspectives can be brought to viewing the genome, in order to create a situation of transparency about transparency:

(1) Scientific research methods describe an individual's genome as containing mutations, that are potentially relevant in the aetiology of disease. We all carry an unexpectedly high number of such mutations even though we are still healthy.[9] Most people carry one or two recessive mutations that can cause early death or infertility in their offspring, one recent study has revealed.[10] The relevance of a mutation seems to depend not only on the effect on the protein that is produced from the gene concerned but also on the genetic

---

[8] Dreyer/Erdmann/Rehmann-Sutter, 2014.
[9] Ashley, 2010.
[10] Gao et al., 2015; Rogers, 2015.

background, i.e. on the other parts of the genome that previously had not actually been considered. Knowledge about an individual genome has become affordable for many, since prices for a whole genome sequence, or an exome sequence, are tumbling. But still, the work that is needed to discover a genome, with its mutations, is extremely complicated. Genetic knowledge is not already present within the genome, waiting to be found and read by us. The genome is not a book with readable letters in it, even though a powerful image of the genome describes it as such: the 'book of life' or the 'instruction book' of the human body. But these words and similar ones are problematic metaphors. A genomic sequence is generated scientifically in a complex process of producing and interpreting data. Sophisticated laboratory devices and computers are involved. It therefore differs from other kinds of knowledge that we are used to dealing with in our lifeworld and everyday experience, due to the ways in which it has originated. In order to understand genetic knowledge it is crucial to take these processes of data generation into consideration. They influence what kind of 'knowledge' this genetic knowledge is (i.e. its epistemology), and how it can become practically significant in our lives.

(2) A phenomenological perspective is different from the scientific factuality of genetic knowledge. Humans are beings who differentiate between themselves as subjects and things as objects. Whereas the genetic sciences treat humans as objects, a phenomenological view focuses on subjective, lived experiences and feelings. Humans have a personal view of their own genetic structure. This subjective view is informed by scientific knowledge but cannot be reduced to it. Questions arise about how to deal personally with this genetic information, and also with the genetic information one could potentially acquire.[11] Genetic information in its present and future-related dimensions is not only new to many but has some interesting particularities and differences compared to other diagnostic information. In addition, the probabilities that are part of the scientific information about genes and their mutations gain personal significance that has to do with lifeplans, with fear and hope, with responsibilities and relationships, with identity and self-concepts,

---

[11] The 'lived genome' is a research perspective. See Rehmann-Sutter/Mahr, 2016.

with narratives about who we hope to be etc., none of which can be explained by probability alone. To be a healthy patient (who could become ill) is a complex experience. Genetic knowledge prompts new dimensions of self-experience that dissolve the traditional experiences of being ill or healthy.

(3) The meanings that are attributed to genetic knowledge in the phenomenological perspective also shape the meaning of the ethical dilemmas people face alongside the potential of acquiring genetic information about themselves or other family members. They are placed in intersubjective relationships. Because it is inherited and appears in the context of a family tree, genetic knowledge always has intersubjective dimensions, which are important to take into account when pondering the ethical issues of genetic testing. It affects not only 'the self' but also 'the other', particular concrete others whom we know or get to know. This fundamentally relational feature of genetic knowledge shapes the questions of genetic responsibility and how morally to deal with genetic knowledge. The higher the impact of genetic knowledge on our own wellbeing, the greater is the responsibility we have in dealing with this knowledge. Do we have a moral duty to inform our children about a genetic risk that we happen to have detected in our own bodies, or would it be morally impermissible to share such knowledge without the free choice of those affected by it? Which moral standards can we use to clarify the ethics of 'disclosure dilemmas'[12] within families? Which moral standards regulate communication between healthcare professionals and clients in counselling situations? And how should studies that include whole-genome analyses be described in the public sphere without raising false expectations?

(4) These questions are in many ways related to socio-political issues such as legislation, discourse and governance. Rules need to be devised that regulate the production and communication of personal genomic knowledge. Transparency both means disclosure under certain conditions and non-disclosure under others. The parties who may legitimately claim access to somebody's genomic information need to be defined. Measures of privacy protection need to be installed and the standards of genetic counselling need to be clarified.

---

[12] Rehmann-Sutter/Müller, 2009.

In order to discuss these issues properly we need the methodologies of political sciences and sociology. They illuminate the collective dimension of dealing with genomic knowledge. Legal systems need to be adapted; the functioning of collective actors such as healthcare systems, insurance and other interested private companies need to be defined.

(5) In all these regards, terminological questions arise that need conceptual philosophical work. We can use terms in everyday language (such as 'human', 'normal', 'disease', 'duty') successfully without always having a clear, nuanced and shared concept of their meaning. Only when we start to analyse these terms do we discover fine differences in their use that might be significant if we only look closely enough. In an investigation of the practical implications of human genomics it is sometimes crucially important to determine precisely who we think are the carrier of genetic information and how we understand these carriers to be subjects of responsibility. Philosophical anthropology and the philosophy of biology discuss central concepts such as 'human' or 'genotype'. The history of these concepts also matters on a meta-level, in order to examine the language that we use to discuss the topics scientifically, phenomenologically, ethically or sociologically. Theoretical philosophy needs to be integrated in transdisciplinary work and can contribute by identifying and constructively criticising key terms of the discourse.

***

The seven chapters of the book do not strictly follow this order. The first chapter introduces the metaphor of transparency, both in general and in its application in genetics. The 'genes' of persons that can become transparent today are not directly observable things in nature but explanatory factors that are associated in a special and non-trivial way with the DNA sequence. How this relation can be understood will be discussed. The second chapter describes the investigative process of DNA sequencing in genomics that leads to the present forms of 'genetic knowledge'. As elsewhere, the methods determine the kind of knowledge that can be derived from experience. At the same time, the metaphor of visibility is discussed with regard to sequence mutations and to sequencing DNA. In the third chapter, anthropological questions are raised. The question is 'who' is the human being that carries genetic information. A particular question arises about human individuality connected to an individual genome. If alterations of the inheritable genome should be avoided, does that include the mitochondrial genome as well? Mitochondria in cells do not contribute to the features that make human bodies different but are crucially important for metabolism, and mutations in this genome can have fatal effects. Chapter four explores the present landscape of personal genomics. Direct-to-consumer genetic testing and genomics services need to be regulated in some way. The chapter uses the example of the proposals of the Danish Ethics Council to clarify key normative concepts that arise in the discussion of governing personal genomics ethically. The ideology of direct-to-consumer genetics and the parts of the moral order with which they interfere will be discussed, and two models of governance are distinguished with regard to both predictive medicine and pharmacogenomics. The spread of genetic information leads to questions about criteria for the good usage of this information. The ethical issues of genetic disclosure are the topic of chapter five. Genetic transparency and genetic privacy may conflict at times, even though they are not mutually exclusive. This chapter therefore discusses aspects of these concepts for the legal and ethical consideration of genetic testing in humans. Topics include the right to know and the right not to know, access to testing, data confidentiality, autonomy and wellbeing. Chapter six opens up a regulatory perspective on genomic investigations. The chapter considers two different case studies of the application of different governance models to the genomic context. The first looks at the expansion of newborn screening (NBS) programmes in several post-Communist Eastern European countries, and examines the policy issues and the influence of political context on the scope and content of genetic and genomic research programmes. The second case study takes an in-depth look at the relationship between private rules and state law in the self-regulation of research around whole-genome sequencing. The scientific committee of the interdisciplinary research group EURAT in Heidelberg, Germany has adopted a Code of Conduct for Whole Genome Sequencing 2013. The discussion of these two cases contributes to responding to the host of challenges of regulation of new medical technologies, which are evolving rapidly. The final chapter treats ethical issues in the communication of genetic transparency. What can be known about a person's genes, by whom and for what purpose? This will be discussed in three specific contexts: non-invasive prenatal testing (NIPT), the pursuit of genetic diagnosis by parents of children with purported or potential genetic conditions, and genetic testing for a known cancer predisposition syndrome. It concludes by highlighting questions for further ethical consideration.

***

This book has been cooperatively written by a group of 18 scholars, largely from different disciplinary backgrounds ranging across molecular genetics, genetic counselling, anthropology, sociology, history, law, literature, bioethics and philosophy. Each of the chapters is a collaborative work by a group of authors. The book in this shape has a special history, which we will outline here briefly. Its origins go back to a call for submissions of projects for interdisciplinary workshops on topics of ethical, legal and social aspects of modern biomedicine (ELSA Klausurwochen) that was issued by the German Federal Ministry of Education and Research (Bundesministerium für Bildung und Forschung, BMBF). We developed an application, and won a generous grant that funded a one-week workshop with over 20 active participants organised at the University of Lübeck in March 2013, and the book project. The workshop was based on a series of invited papers by senior international scholars and a selection of applications of younger scholars to an open call for papers. The resulting workshop brought together perspectives from six European countries plus Canada. It resulted in a provisional book plan with the seven chapters sketched out. All authors (except one) who have participated in this book project were also present during the workshop and gave earlier versions of their papers that were then substantially reworked within chapter author groups after the workshop, and then incorporated into the coherent text of each chapter. Daniela Appee, Koen Dortmans, Thomas Douglas, Gösta Gantner, Gabriele Gillessen-Kaesbach, Catherine Lyall, Thomas Meitinger and Tim Strom all contributed to the workshop but were unfortunately unable to participate in the book project afterwards; Benedikt Reiz joined the book writing group after the workshop.

We thank all the participants in the workshop and the book writing process for rich discussions and a wonderful working atmosphere, and also for their patience -- which was much needed during the writing and finishing process of roughly 3 years. We thank the Institute for History of Medicine and Science Studies and its director, Cornelius Borck, for generously hosting the workshop. We thank Christina Schües and Wolfgang Lieb for joining us in an evening public panel discussion on the ethics of personal genomics and sequencing during the otherwise closed workshop; Kathrin Langkau, Brita Dufeu and Angela Mötsch for secretarial assistance; Martina Steinig for support in the university accounts department; and Simone Mistry and Matthias von Witsch at the National Aeronautics and Space Research Centre of the Federal Republic of Germany (DLR) for their administrative support in managing the BMBF grant. We thank Monica Buckland and Jackie Leach Scully for revising the English text and for many helpful suggestions that have led to clarification of details. Heartfelt thanks to Lisa-Marie Müller for her immense work on the layout, in copy-editing and supervising the correction of the proofs. Bill Johncocks has skilfully produced the index. We are grateful to the series editors Peter Derkx, Erica Haimes and Harry Kunneman for accepting our idea for their book series, and Eric van Broekhuizen at Brill Rodopi for his uncomplicated and efficient support. We acknowledge funding from the German Federal Ministry of Education and Research (BMBF).

Literature

Ashley E.A. et al. Clinical assessment incorporating a personal genome, Lancet 375(9725) (2010), 1525- 1535.

Brin, David. The Transparent Society. Will Technology Force Us to Choose Between Privacy and Freedom?, Perseus, Reading, 1998.

Cadwalladr, Carole. What happened when I had my genome sequenced, The Guardian, 8 June 2013.

Gao, Ziyue, Waggoner, Darrel, Stephens, Matthew, Ober, Carole, Przeworski, Molly. An Estimate of the Average Number of Recessive Lethal Mutations Carried by Humans, Genetics 1999 (2015), 1243-1254.

Gymrek, Melissa, McGuire, Amy, Golan, David, Halperin, Eran, Erlich, Yaniv. Identifying Personal Genomes by Surname Inference, Science 339 (2013), 321-324.

Hayden, Erica Check. Privacy Protection: The genome hacker, Nature 497 (2013), 172-174.

Pasquale, Frank: The black box society. The secret algorithms that control money and information, Havard University Press, Cambridge, 2015.

Rehmann-Sutter, Christoph, Mahr, Dominik. The Lived Genome. In: Whitehead, A., Woods, A., Atkinson, S., Macnaughton, J., Richards, J. (eds.). Edinburgh

Companion to the Critical Medical Humanities, Edinburgh University Press, Edinburgh, forthcoming 2016.
Rogers, Nala. Genomes Carry a Heavy Burden, Nature News 15 April 2015. doi:10.1038/nature.2015.17304.
Rogith, Deevakar, et al. Attitudes regarding privacy of genomic information in personalized cancer therapy, Journal of the American Medical Informatics Association 21/e2 (2014).
Schües, Christina. Wagnis Zukunft: Braucht Vertrauen Transparenz? In: Hirsch, A., Bojanić, P., Radinković, Ž. (eds). Vertrauen und Transparenz - für ein neues Europa, Institut für Philosphie und Gesellschafttheorie, Belgrade, 2013, 56-80.
Wible, Brad, Jasny, Barbara (eds.). The End of Privacy, Science 347 (Special issue) (2015), 491-514.

1

The Idea of 'Genes' and Their 'Transparency'

Christoph Rehmann-Sutter and Malte Dreyer

Introduction

In an opinion piece in the 9 October 2013 issue of the journal *Nature*, Harvard geneticist George Church asked "why so few people are opting to inspect their genome". But how could they? How could somebody inspect the genome? To 'inspect' must mean something different here than to inspect, for instance, one's skin or teeth. More intimate parts of our skin that we might have inspected by a dermatologist are normally covered by clothes. Teeth are visible, even if they are hidden some of the time by our lips. When we make somebody laugh we have a moment to inspect her or his teeth, which would actually be considered rude, because the meaning of a laugh is to express joy, not to show one's teeth to another person. A genome, however, is not something that we can so easily look at. We cannot so easily trick other people into showing us their genes.

Church does not mean that we should take one of our cells, put it under a microscope and look at the chromosomes inside. What he means is rather this: "why have so few obtained and interpreted their own genome sequence?".[1] The activity of inspection here is obtaining a genome sequence and having it interpreted. This implies three things: first, that the genome becomes visible as a sequence. It is not the shape of the macromolecular complex of chromatin that we are interested in but the sequential order of four different types of nucleotides (G, C, T, A) that are lined up in this thread-like molecule of DNA. The genome is the sequence of the molecule, its information content, not the molecule in its corporeality. The second

[1] Church, 2013, 143.

thing implied in this proposition is that we cannot do the job of seeing ourselves. We need somebody else to do it for us, because sequencing is a terribly complicated procedure that needs costly laboratory equipment and know-how (see the next chapter). If we decide to inspect our genome we rely on the people who can deliver our sequence to us. But even then, the meaning of this sequence does not reveal itself to us. It needs to be interpreted, and the job of interpretation again needs highly specialized knowledge. We need professional genetic interpreters (genetic counsellors) to read the sequence to us in an understandable way. Again, we rely on their interpretation. We might have our own ideas, and after hearing what the specialists say we transform that knowledge and integrate it into our lifeworld.[2] Thus, there is quite a lot of dialectics hidden in this idea of 'inspecting' one's genome. And dialectics is also hidden in our book title, the 'transparency' of genomes. Visibility, inspection and transparency are all metaphors, and their meaning in the genetic context is far from trivial.

This chapter intends to unpack some of these dialectics. In doing this we also need to talk about the contraries: the invisibility, the deceptions or illusions that could be connected to the exposure of personal genomes, and the opacity (intransparency) of the genome to people. This chapter explores the dialectic that emerges from two different terminological contexts: the gene, the genotype and the genome on the one hand, and the family of visibility metaphors on the other. We will trace a few important points of their genealogy in the genetics discourse of the 20th and 21st centuries. But before we start we need to discuss metaphors as such: as a fascinating linguistic and rhetorical phenomenon. What is their logic, what is the grammar of metaphors?

The grammar of metaphoric speech

The most well-known conventional definition of a metaphor was suggested by Aristotle: a metaphor is an abridged comparison.[3] By

---

[2] Rehmann-Sutter/Mahr, 2016.

[3] Aristotle, *Rhetorics* 1406b 20ff.

saying, for example: 'I show you your genome', we compare a complex set of acts, which includes collecting body material, extraction of DNA, sequencing and interpreting, and is difficult to imagine – even for those who know how to do it technically. We compare it with something that we all know much better from everyday life: showing things to other people, so they can see them. By comparing genome research with acts of showing, we explain something. We bring meaning from one field into another field, in order to make the latter meaningful. We relate two domains to each other. Contemporary linguists call them the 'source domain' and the 'target domain'[4], and they say that the comparison is an act of 'mapping' between these two domains. Cognitive linguists[5] emphasize that the role of metaphors is to bridge normally independent domains of cognition: showing things to each other is a cognitive act in everyday life, and genetic investigation is a cognitive act in science. The everyday experience of showing and seeing is the source domain and genetic investigation is the target domain. Sociologists of knowledge[6] emphasize the role of metaphors as carriers of knowledge and meaning across discourses and disciplines. This explanation of the metaphorical process is perhaps more pertinent when we talk about what is shown. We may ask, "What are you showing me when you show me my genome?" One answer is: "Your genome is your blueprint, a sort of plan that told your body how to build you." Then we have two discursive domains and a transfer of knowledge and meaning. The domain of genetics is infused with meaning that is rooted in the practice of building and engineering, where things are built according to blueprints and plans.

The German language media have frequently used the metaphor *'der gläserne Mensch'* (the glass or transparent person) in order to point out the increasing transparency of human beings that comes with genomics: the human body becomes transparent as if it were made of glass. You can look at the insides, its hidden parts, its constitutive information, contained in its genes. The term 'human' is related to another term: 'glass'. The human made of glass is a metaphor

[4] Hellsten, 2008, 185.
[5] Lakoff/Johnson, 1980.
[6] Maasen/Weingart, 2000.

according to the classic Aristotelian definition. A target domain is explained with a comparative domain by identifying it with something: in genetics, humans become glass objects. They become crystal clear. Others cannot see through them (like through a window); that is not what is suggested by this comparison. But others can look into them (like into a bottle) and see what is in them: their hidden potentials and also their hidden weaknesses. They can see things that the person would prefer not to show. Genetic information is therefore, at least in some legislations, considered to be personal information that nobody is allowed to collect or to use without the consent of the person to whom the information relates. She or he needs to consent (see chapter 5 below). The comparison of humans with glass also makes them appear fragile. They become vulnerable in new ways. But the comparison of human bodies with glass bodies also makes humans appear hard and not malleable, perhaps cold as well. The genes are given, inherited. We cannot change them by using force. If we try to do that they may shatter. So this comparison implies that humans must be treated with special care. You could even go so far as to say: as a metaphor it does not seem to work at all. Genomic knowledge does not make the person more transparent at all – certainly not in the visual sense that talking about 'glass' would imply. At least to non-geneticists it makes it less transparent, because less immediately comprehendable.[7] – The target domain 'human' is endowed with a bundle of different meanings, which only through that comparison become plausible in a certain way and create a certain kind of evidence and raise certain questions. Metaphors bundle particular properties by using a graphic comparison, an image.

In order to work in this way, a linguistic image must fulfil some logical requirements. Knowing these requirements is important both for the theory of metaphor in general and to understand the idea of 'genetic transparency' in particular. On the one hand, the source and target domains must be sufficiently different to motivate us to look for possible common features. Most competent speakers would consider the sentence 'Humans are living beings' not as metaphorical speech but rather as a factual statement about the term 'humans'. In order to count as metaphorical, a way of speaking must map features

---

[7] We are grateful to Jackie Leach Scully for this point.

of one domain onto a domain of a different kind. At the same time the domains must have something which is similar. If no comparison is possible, there is no basis for a metaphorical statement. An example of too large a difference would be the sentence: 'Humans are a 3', since no speaker can bring forward features which are shared by humans and the number three. The sentence is absurd.

If a transfer of features is possible, similarities become visible. We cannot therefore compare (in a meaningful way) everything with everything. Aristotle called such a transferable feature 'tertium comparationis'.[8] With regard to the explanation of the conditions of the 'human' being in the world, the 'tertium' varies between cultures and times. In a hypothetical culture that exclusively knows *black* glass, the image of being transparent and, in all relevant aspects, open to inspection would not be generated by using the comparison with glass. Air or water would perhaps be better comparators. In a culture that excludes humans from the category of living organisms (because for example humans are seen as transcendent beings, or the term 'living organism' is reserved for animals without speech), the sentence 'Humans are living beings' could indeed be interpreted as metaphorical. The question of whether a trope is metaphorical or not, and what being metaphorical signifies, is therefore decided in a process of interpretation that depends on culturally specific habits of discernment and distinction. The concept of a metaphor that provides meaning therefore not only includes the logical structure that we have tried to describe, but also a cognitive activity that is framed by cultural contexts.

The metaphor of 'genetic transparency' is a good example for this, since it refers to the possibilities and limits, the benefits and risks of genetic knowledge. This knowledge is quite young and exceptionally dynamic. It enters the interpretation of the phrase 'genetic transparency'. What it can mean depends on the meaning of the term 'genetic', which we will look at more closely below. Let us however go back one step and ask how this metaphor works. Transparency is a feature of objects that are visible. Are genes visible in the way glass is visible, for instance? Transparency, in the phrase 'genetic transparency', can mean either the transparency of genes or the

---

[8] Aristotle, *Poetics* 1457a, 30 ff.

transparency of the carrier of these genes. In the latter sense, transparency would signify a human being whose genetic features are known and open for interpretation. If we take a look at the transparency metaphor, we notice that in most cases 'transparency' refers to the carrier of the genes as a "person" and not merely to the pure organic material. This is why the practice that makes a human being genetically transparent raises normative questions. At the same time 'transparency' qualifies an extraordinary perception of something or somebody. As a result, the metaphor 'transparency' has implications for the observed object as well as for the standpoint from which this object is seen. In general this metaphor is characterised by these two-sided meanings. This ambiguity is relevant in several respects:

(1) I cannot share my gaze with anybody else. Seeing is a practice where the specific content is distinctly bound to a standpoint, a 'perspective'. This is not, or not in the same way, the case for other kinds of sensory perception, such as smelling or tasting. This feature of sight makes words that originate in the semantic field of seeing especially attractive for their metaphorical use in epistemological contexts.

(2) Standpoint epistemology can also be applied to normative assessments. Transparency of genetic features might be valuable from one perspective but not from the other. Our salary statements as average academics in Germany should be transparent to 'us'. Others might have an illegitimate interest in it in certain circumstances (competition, academic reforms etc.) but most people would just not care how much we earn. The same is true for genetic transparency: the value of obtaining access depends on the circumstances and on the perspective.

(3) In certain contexts, transparency is a positive value, in others it is an issue. We speak of a free and fair election only if the counting of votes and the communication from the election offices to the government is transparent. However, every voting decision should be invisible to others, and voters are given curtains to shield them as they make their marks on the ballot. Therefore it is important to clarify beforehand which aspects of a practice should be transparent to whom. This itself should be based in transparent arguments. Hence, the rules of genetic transparency and genetic opacity are truly

political. These rules of transparency need to be decided in a fair and open participative process.[9]

(4) The transparency metaphor emphasises the exclusivity of standpoint and perspective of a subject. If we focus not on the stem 'parere' but on the prefix 'trans', it becomes evident that we capture transparent objects completely, totally, and from all sides, not just partially. A transparent object is visible from all sides. I see it just like others who see it from another perspective, because I can perceive the front side and the obverse as well. Therefore we not only need to consider the perspective from which something becomes visible, but also whether we want to show something at all. If it is shown transparently, it becomes visible totally, from all sides and in a shared perspective. The transparency metaphor has a holistic implication: we see through all of it.

The combination of 'genetic' and 'transparency' in the first place seems obvious: if DNA sequences become accessible and in a way 'visible', our bodies become (metaphorically) transparent with regard to genetic factors, and we can speak of 'genetic transparency'. Genetic transparency therefore highlights possibilities of access and disclosure, knowledge and ignorance, understanding (or misreading) personal genetic information. All possibilities of creating transparency, since they can relate many individuals to each other, or set them against each other, in many ways, are not just individual actions but characterize essentially social acts and political relationships. The knower knows about somebody else's 'genes' and this gives her or him power or authority in some way. There are knowers, or people who pretend to know, and there are people whose information is known. This subject-object distinction raises the question of to whom personal genetic information is, or should be, transparent; which information is, or should be, conveyed; who decides, or should decide, about disclosure; in what way genetic information becomes, or should become, accessible, and so forth. Let us now look more closely at the ideas that are connected to the words 'gene', 'genetic', 'genotype' and 'genome'.

---

[9] Dreyer/Erdmann/Rehmann-Sutter, 2014; Rehmann-Sutter/Mahr, 2016.

A very short biography of the genes

The language of biology is very rich in metaphors. In this discourse we encounter 'cells' (in analogy to monks' or prisoners' cells?), genetic 'programs' (similar to computer programs but dissimilar to concert programmes or radio programmes?), or 'codes' (compared with Morse code or codes of conduct?). Genes are supposed to be 'selfish', if we believe Richard Dawkins' formulation,[10] but this should not mean that the people who are made of selfish genes are selfish. They could well be made altruistic for selfish reasons, if this is a strategy for transmitting their genes more efficiently into the following generations.

The philosopher Paul Ricoeur distinguished between living and dead metaphors. A metaphor is alive, as long as in pronouncing in the current context, we still can feel a certain resistance: a resistance against *this* use of the word. If we can still feel a resistance when using it in this sense, the metaphor is a living one; if we just use the word without feeling anything uncommon, according to Ricoeur the metaphor is dead. This of course can change in the history of a discourse. Metaphors can die. Whether a metaphor is dead or alive therefore depends on the use of language in a specific cultural context.[11]

The use of metaphor in biology does not imply that scientific research is less anchored in reality. Rather, metaphorical speech interacts dynamically with propositional language in a productive way. This interaction causes terminological innovations, as Stanley Cavell (1979) has stressed.[12] Many terms in scientific language have their roots in a living metaphor that has become a dead metaphor. In order to avoid misunderstandings it is sometimes important to remember the origin of a term and the reason why it was expressed in pictorial language in the first place. Further, understanding a metaphor's meaning requires knowing the purpose for which a metaphor was

---

[10] Dawkins, 1976.

[11] Ricoeur, 1975.

[12] Cavell noted that transfers of meaning, as we use them in metaphorical speech, belong to the fundamental principles of language and thought in general. This principle is also known as "projective imagination".

invented. The term 'cell' is a good example of the development of valid biological terms through the use of metaphors.

For most biologists, the 'cell' is a dead metaphor. The word has lost its resistance when we use it for these little membranous compartments of organisms that contain cytoplasm, organelles and a nucleus, which were what was meant by Matthias Schleiden and Theodor Schwann in 1838 when they developed the cell theory. In today's language, the word 'cell' literally seems to *mean* these things. The monk's or nun's cell may be mentioned in an etymological lexicon. We no longer feel that 'cell' is a metaphor. The term 'genetic program' also seemed to have lost its life for several decades and to have become a dead metaphor. But now it has been revived. Today, many biologists would use the word 'program' only with resistance, and if they do use it then only with the utmost care, knowing about the ambiguity of its meanings, some of them misleading. It is no longer clear that the genome actually works in the same way as a computer program.[13] The program is a metaphor that has some wanted and some unwanted implications, and today it is again *felt* to be a metaphor.

What about the gene? The gene when used in biomedical language[14] is no metaphor at all. It is a technical term that has been introduced to signify a particular thing. What is it? Many people today would simply say a gene is a stretch of the DNA molecule, or it is a section of its base pair sequence, or a spot on a chromosome. Genes are important because they function as a predisposition for an embodied characteristic, for instance the predisposition for having blue or brown eyes, long or short fingers, or for the risk of developing a disease such as Parkinson's. The term 'gene' is however much older than knowledge of DNA, the double helix or DNA sequences. It was coined in 1909. Wilhelm Johannsen, a Danish botanist, was interested in replacing both the existing ambiguous German term *Anlage* ('disposition') and Charles Darwin's artificial word 'pangene' with a more precise term that could be defined and scientifically useful. He wrote: "Merely the simple idea should be

---

[13] Keller, 2000.

[14] It can be used metaphorically in other contexts, for instance when a football club claims before a game that "it is in our genes to be winners".

expressed that a property of the developing organism is or may be conditioned or co-determined, through 'something' in the gametes. No hypothesis about the nature of this 'something' should be established or defended."[15] This was in many respects a remarkable definition.

It was intended to express a simple idea, nothing complicated. The simple idea that Johannsen wanted to capture was that the properties of organisms, which develop throughout their lives, are *somehow* conditioned or co-determined by *something* in the egg and/or sperm that are combined in fertilization. So the word 'gene' does not introduce the idea that a certain molecule or a certain substance does this determination during the development of the organism, but it does introduce a causal relation between 'something' that can be seen and experienced – a 'phenotypical' property, Johannsen said – and 'something' that must have been present in the gametes, the 'genotype'. The relation is one of conditioning or co-determination. The other source of causal influences in this co-determination relationship is the environment. The etymology of the terms 'gene' and 'genome' relates back to the Greek verb *gignomai* – to come into being, to be born. Johannsen derived the more handy word 'gene' from 'genotype', which he had coined a few years earlier in 1906. Genes are related to traits, and can be used in the plural in the same way as traits can be plural. 'Genetics' then started to study this relationship and it became the name of a scientific discipline.

The freedom from any material hypothesis certainly contributed to the success of the term 'gene' in the history of genetics. Like an empty picture frame it has been filled with different things as potential genes. Through the history of genetics the things that were seen to represent the genes changed, but the term 'gene' survived. At first the genes were the Mendelian *Faktoren*, then they became loci on chromosome maps. Before the experiments of Oswald Avery in the 1940s, most scientists believed that genes must be proteins in some way. After the successful work of Franklin, Watson and Crick in

---

[15] Johannsen, 1909, 124: "Bloss die einfache Vorstellung soll Ausdruck finden, dass durch 'etwas' in den Gameten eine Eigenschaft des sich entwickelnden Organismus bedingt oder mitbestimmt wird oder werden kann. Keine Hypothese über das Wesen dieses 'Etwas' sollte dabei aufgestellt oder gestützt werden.".

1953, genes became stretches of DNA. More recent authors have suggested seeing the genes as interactive processes: those processes that lead to one polypeptide.[16] Some even include non-DNA factors as genes: 'epigenes' also contribute to heredity. Geneticist Raphael Falk summarizes more than 100 years of scientific history by asking: "Were genes hypothetical constructs, autonomous structural entities or loci of differential functional emphases along an integral chromosome?"[17]

The word 'gene' is not and never has been a fixed entity. It is a living term (but no metaphor), which has been used with a whole series of different meanings. Why do we not believe that the gene is a metaphor? The word was new in 1909 and it had no common use before. That's why. There is no source domain or target domain, which would be necessary for metaphorical speech, where words are used in a figurative sense. As mentioned, metaphorical use *of* the word 'gene' is still possible: for instance, when somebody jokingly abdicates responsibility for his idiosyncratic behaviour, because it is 'genetic'. This, however, had no lasting impact on the current usage of the word 'gene'.

The life of the word 'gene' comes from somewhere else: it is alive because its objective signification has changed much over time. Current research reveals that, in order to acquire the function of 'genes', sequences of DNA need an interactive context, a cellular system that is specific for tissues and cell lines. Therefore, a stretch of DNA cannot be a gene without having many other things in place as well. Here is a quote from Kunihiko Kaneko, who writes from a systems biology perspective: "... [W]e should be studying models of interactive dynamics. Then, we should inquire whether, within such dynamics, the asymmetric relation between two molecules is generated so that one plays a more controlling role and therefore can be regarded as the bearer of genetic information."[18] There are no molecules that, by their very nature, assume a controlling role. This would be an essentialist assumption, which is not warranted by scientific evidence. Hence, the context is necessary for the proper

---

[16] Griffiths/Neumann-Held, 1999.

[17] Falk, 2009, 128.

[18] Kaneko, 2006, 20.

function of DNA *as a gene*, i.e. as a causal factor for a certain phenotypic trait. We will now explain this in more detail.

Johannsen's definition of the term 'gene' was formal. The genes were hypothetical determinants of the organism's characteristics. The definition *had* to be free from any material hypothesis, in order to explain this causal relationship, which was connected to the hypothesis. See again the very exact way in which he wrote this: "The word 'gene' is completely free from any hypotheses; it expresses only the evident fact that, in any case, many characteristics of the organisms are specified in the gametes by means of special conditions, foundations and determiners which are present in unique, separate and thereby independent ways – in short, precisely what we wish to call 'genes'."[19] It is important to note that the freedom of hypothesis, which has frequently been emphasized as the secret of the power of the word 'gene' in the 20th century, includes more than just the chemical nature of the genes, i.e. whether proteins or nucleic acids assume this role in the cells. It also includes also the *form* of involvement in a causal relationship: As he has put it, it could be "special conditions, foundations and determiners". Genes might be molecules indeed, but they might also be processes, or conditional networks of interactive dynamics – to take up Griffiths', Neumann-Held's or Kaneko's terms. Epigenetic factors, many of which have been studied in detail (the DNA methylation patterns and other inheritable factors), according to Johannsen's definition, can also assume the role of a 'gene'.

But in another respect Johannsen's definition has been modified. Genes are not considered to be in gametes alone. Bacteria, the genetics of which have been extensively researched, do not have gametes at all. They just divide. And of course scientists have investigated their genes and genomes, which is their DNA. There is also a genetics of viruses. With the rise of developmental genetics, scientists' interest also focuses on inheritance between different cell generations within one and the same multicellular body. The genetic relationship is therefore not bound to the link between gametes and the new organism in cases of sexual reproduction. It has a broader meaning and refers to many kinds of generative processes: cells

---

[19] Johannsen, 1909, 124.

dividing and differentiating, viruses multiplying in host cells, and *a*sexual reproduction as well, as is normal in many species, such as the polyp *Hydra*, at least under favourable conditions. When local conditions deteriorate, *Hydra* may be induced to enter sexual reproduction; so-called I cells differentiate into gametes. Between each sexual generation an indeterminate number of asexual iterations may occur.[20]

But it is a different semantic layer of the term 'gene' that is predominantly responsible for the striking aliveness of this term: in a piece about the biography of the gene, it is therefore appropriate to talk about the *symbolic* level of genetic explanation.[21] In our philosophical reflections on molecular biology, the genes are seen not only as the physical entities they are, and the term 'gene' is understood not only as an extremely powerful heuristic and successful explanatory strategy. It is also part of a *metaphysics* of the living.

This might not have been intended by Johannsen himself. But it was certainly meant by Max Delbrück, one of the founders of classic molecular biology in the 1920s, who, in a reflective piece about the impact of molecular biology written later in his life, famously related DNA to Aristotle's *eidos*:[22] "It is my contention that Aristotle's principle of the 'unmoved mover' perfectly describes DNA: it acts, creates form and development, and is not changed in the process." The unmoved mover, however, was nothing less than Aristotle's characterisation of God in relation to the universe. This same relationship, according to Delbrück, exists between the DNA and the organism. For many philosophical interpreters in the 1970s it must have been obvious that the discovery of the helical structure of DNA by Rosalind Franklin, Francis Crick and James Watson in 1953, and later of the basic mechanisms of DNA replication and of its coding function for mRNA and amino acids, had deep ontological implications.[23] Ernst Mayr[24], one of the leading contemporary

---

[20] Buss, 1987, 16f.

[21] The relevance of symbols for our self-understanding is discussed below in chapter 3.

[22] Delbrück, 1971.

[23] see Engels, 1982, 81; Rehmann-Sutter, 1993, 94.

[24] Mayr, 1976, 400f.

interpreters, wrote: "Just as the blueprint used by the builder determines the form of a house, so does the eidos (in the Aristotelian definition) give the form to the developing organism, and the eidos reflects the terminal telos of the fullgrown individual. [...] There is, of course, one major difference between Aristotle's interpretation and the modern one. Aristotle could not actually see the formgiving principle (which, after all, was not fully understood until 1953) and assumed therefore that it had to be something immaterial. [...] Since the modern scientist does not actually 'see' the genetic program of DNA either, it is for him just as invisible for all practical purposes as it was for Aristotle. Its existence is inferred, as it was by Aristotle." A few years later, Mayr wrote: "Aristotle's eidos is a teleonomic principle which performed in Aristotle's thinking precisely what the genetic program of the modern biologist performs."[25] Genetic information, contained in the sequence of DNA, seemed to be exactly what Aristotle meant by the form-principle: what structures the matter (*hyle*) and guides the shape, the appearance and all the perceptible characteristics of all things during their development from the inside is the idea (*eidos*) of this thing and for living things their soul (*psyche*).[26] Molecular biology, so it seemed, vindicated Aristotelian metaphysics. The DNA was considered to contain the information that tells the bodies of living organisms who and what they are.

This interpretation of molecular biology is obviously essentialist. Not all biologists who were active in research shared this essentialism. But it certainly formed the public understanding of genetics.[27] How could that happen? The functional principles of DNA were quite an intricate matter, not easily accessible to those not acquainted with basic biochemical and biophysical knowledge. Our answer to the question of how it could happen is this: it did not matter so much how DNA works in the cells as a molecule. It was more important to have an accessible *image*. In the public sphere the image explains how to imagine the activity of the genes. This image was – the *genetic program*. The genetic program is perhaps one of the most significant innovations of the scientific language of the 20th

---

[25] Mayr, 1982, 88.

[26] see Aristotle, Metaphysik VII, 1035b14-16.

[27] Nelkin/Lindee, 1996; Turney, 2009.

century. In the quotes from Ernst Mayr, this term appears prominently. The introduction of the term around 1960 was a key symbolic move. The 'genetic program' provided a speculative but scientifically reasonable answer to the question of how the genome can direct the development of an organism, without – and this was important – claiming goal-directedness, or teleology. Teleology has been considered theological, metaphysical or ideological. The claim that natural processes strive to reach an end is a non-scientific idea.

Lily Kay has extensively investigated the history of the discourse of information in genetics. She has found that the term 'genetic program' appeared for the first time in 1959, in a notebook of Jacques Monod, a leading microbiologist at the Institut Pasteur in Paris. He sketched out a conceptual way to reconcile the 'necessity' of endpoint-directed processes with the 'hazard' of all innovation in the evolution of life. A solution seemed to him to be contained in the mechanical function of a computer-like 'genetic program'. A program brings about *seemingly* goal-directed behaviour of a system purely by the principles of mechanical function. We all know this when we use personal computers. When we boot it up, the operating system tends to bring the computer back into the proper shape we can work with. But we know that the operating system is not thinking. It does not have intentions. It is pure information technology, which is in a way pure mechanics.

In published writings, the term 'genetic program' appears for the first time in 1961 – twice. At the end of a joint review article about the regulatory mechanisms of protein synthesis Jacques Monod and François Jacob wrote: "[T]he genome is considered as a mosaic of independent molecular blueprints for the building of individual cellular constituents. In the execution of these plans, however, co-ordination is evidently of absolute survival value. The discovery of regulator and operator genes, and of repressive regulation of the activity of structural genes, reveals that the genome contains not only a series of blue-prints, but a co-ordinated program of protein synthesis and the means of controlling its execution."[28] The 'blue-prints' are the coding sequences for proteins. The activity of these 'structural genes', as Jacob and Monod called them, is controlled by a

---

[28] Jacob/Monod, 1961, 354.

set of 'regulatory genes', which together form a program of protein synthesis during the lifetime of the cell. In the same year, a lecture by Ernst Mayr was published in *Science*, where the term 'genetic program' assumed the leading role. It was a theoretical treatment of the teleology problem in biology, as we have just described it. Mayr developed a theoretical solution: systems that work on the basis of an informational code can have the features of goal-directed development and behaviour. With this move, the program metaphor was definitely entered into the discourse. In the following years it was woven into a rather extended network of other figurative terms taken from the field of text, books, plans and instructions.[29]

The common feature of these attempts to capture the philosophical basis of genetics is that they generalize. They generalize the coding principle that biologists used to explain the relationship between genetic sequences and protein sequences to hypothetically explain the entire genome-organism relation. Jacob and Monod's 'blueprints' were coding sequences for proteins, and to take just one representative author, Robert Shapiro, a few years before the start of the Human Genome Project indicated the significance of the genome with his eyecatching book title *The Human Blueprint*, which was reinforced by the subtitle: "The race to unlock the secrets of our genetic script."[30] Now it is not a protein molecule that is built after genetic blueprints but the whole human body. The genome contains 'our script'. This is a much more grandiose claim. The genetic script should carry the secret of human existence, and it was to be revealed by uncovering the secret sequences of DNA.

The sociologist of prenatal genetics, Barbara Katz Rothman, observed perceptively that genetics has become more than a science. "It is a way of thinking, an ideology. We're coming to see life through a 'prism of heritability,' a 'discourse of gene action,' a genetics frame. Genetics is the single best explanation, the most comprehensive theory since God."[31] What provides genetics with this status is not that it might explain virtually everything that is important in organic life, but that it seems to explain the essence of what it is to be a living

---

[29] Kay, 2000; Rehmann-Sutter, 2003; Peluffo, 2015.
[30] Shapiro, 1991.
[31] Rothman, 1998, 13.

being. This is the genetic lens. It is a reconfiguration of worldviews towards seeing *what is essential* in *what is genetic*. There are of course many more things that are essential in human life than those that could possibly be explained by genetics. But the ideology tends to deflect the light of our perception such that it seems the key to understanding the deepest questions of human existence is genetic information.

The cultural discourse of genetics contains plenty of examples that can be brought as evidence for this. We will mention just one of the most prominent. In June 2000, a working draft of the map of the human genome was ready to be presented to the world. Bill Clinton, then president of the United States, explained in a widely reported speech at a White House press conference what it means for humankind to learn about the human genome: "Today we are learning the language in which God created life. We are gaining ever more awe for the complexity, the beauty, and the wonder of God's most divine and sacred gift."[32] Francis Collins, who was the head of the official US-funded Human Genome Project, speaking after Clinton, responded: "It's a happy day for the world. It is humbling for me, and awe-inspiring, to realize that we have caught the first glimpse of our own instruction book, previously known only to God."[33] What should astonish us about these sentences is less the fact that the most rigorous science goes so smoothly together with a firm belief in a supernatural god. As we know, many eminent scientists were also firm believers and saw no contradiction between the religious and scientific views of the world. More surprising than this is the interpretation of the meaning of genetic information, or more precisely, of the DNA sequence. The Human Genome was just a sequencing project; it entailed no functional interpretation of the genes. The sequence is seen as (i) a meaningful language, (ii) the language that creates life, and (iii) a book that contains the instructions for building, shaping and developing the body and all its functions. The genetic relationship introduced by Johannsen in 1909 as an explanatory relationship between factors and features, and which later was transformed into an explanatory relationship between

---

[32] Collins, 2006, 2.
[33] Collins 2006, 3.

stretches of DNA and proteins, has been generalized into an explanatory relationship between the genome and life, including human existence.

This is the glory of DNA. Interestingly, however, there is an irony in this story. The very success of empirical genetic research undermined the foundations of its own ideology. The evidence gathered since the 1970s and 1980s about the functioning in multicellular organisms during their development from the zygote is at many points at odds with the concept of a genetic program. It was developmental genetics of multicellular organisms such as the fruit fly *Drosophila* or the worm *Caenorhabditis* that revealed many details of the striking *context dependency* of the information that is supposed to be 'contained' in the DNA sequence. In 1977 alternative splicing was discovered. This is the ability of a cell to create more than one protein 'recipe' from one and the same gene – dependent on the state and the context of this cell. Certain stretches of the mRNA copy of a gene are cut out – these bits are called 'introns' – and the remaining stretches (the 'exons') are combined, leading to a mature mRNA molecule that is then used in the cytoplasm as the actual blueprint for the protein. By cutting out different pieces the cell can form different mature mRNAs from one and the same gene. And it has been discovered that the selection, which cutting pattern is to be applied, depends on the situation of the cell. The cutting pattern differs for instance between different tissues of the body, or between different developmental stages of the organism. In this way, the human body can produce hundreds of thousands of different proteins during its lifetime, always at the right time and at the right place, from only about 25,000 genes in the genome. This has a striking consequence: The information of DNA is not self-contained, not autonomous.

A few years later, mRNA editing was discovered.[34] This is an even more surprising mechanism by which cells can intervene in the sequence of an mRNA molecule before using it as a template for protein synthesis. They can add or cut out letters, or change some others. In the first observed case of mRNA editing, in Trypanosome mitochondria, the cell adds four nucleotides which are not encoded in DNA. Since transcription works with a triplet code, the insertion

---

[34] Benne et al., 1986.

of four nucleotides necessarily causes a shift in the reading frame. It therefore changes the meaning of the whole sequence that comes after (downstream of) the insertion, until the end of this mRNA molecule is reached. Because only this modification of RNA results in a functional protein, mRNA editing is a necessary process in the cell. It is not a disruption of previously correct information but rather the formation of the correct information from an immature draft. Therefore the metaphor of 'editing' seemed to be appropriate here: to 'edit' a text (for instance an article in a newspaper) means to eliminate mistakes or difficult sentences, in order to make the text readable and to bring out its true meaning more clearly, the meaning which had supposedly been intended by the author. Editing is not writing a new text.

There are other phenomena of the context-dependence of genetic information, which are difficult to account for within the conception of a genetic program.[35] The program metaphor implies that there is something there on an informational level that functions as a list of instructions for development. But if this information is itself *formed* during development, the assumption of a program does not make sense. It was Susan Oyama who for the first time systematically explored the theoretical implications of the ontogeny of genetic information.[36] Her suggestion is that the organism needs to be reconceptualised as a 'developmental system' that *uses* DNA information together with information from different sources in appropriate ways according to its need during the trajectories of development. The genetic information for development does not pre-exist development.

This is a quite fascinating idea with far reaching consequences. Paul Griffiths and Karola Stotz[37] explain genes within the lines of an Oyamian developmental systems approach as follows: they are "things an organism can do with its genome," – " ways in which cells utilise available template resources to create biomolecules that are needed in a specific place at a specific time." The genes are not given

---

[35] For reviews see Schmidt 2013, Falk 2009, Moss 2003, Moore 2001, Keller 2000, Neumann-Held/Rehmann-Sutter 2006, Rehmann-Sutter 2005.

[36] Oyama, 1985.

[37] Griffiths/Stotz, 2006.

but regularly made by the organisms who use them step by step. This has caused the once relatively simple concept of the gene (one stretch of DNA codes for one protein) to "burst", as French molecular biologist François Gros noted.[38] Genes function in systemic and highly contextual interactions that are only partly understood.

Summarizing this biographical sketch of the gene:[39] During the years of its decoding, the genome has changed its identity. Or, as Jon Turney[40] has put it, it has slipped into an identity crisis. We see a paradigm shift in how genes and genomes have been conceived. The genetic program metaphor, and neo-Aristotelian essentialist gene metaphysics (we can dub it 'genetic essentialism') that had been connected to it, no longer correspond satisfactorily to the currently available scientific evidence. Today, the genome (i.e. the DNA molecule, complemented with different layers of epigenomic information), is rather an *informational resource* that is utilised by the cells to synthesize proteins and RNAs that are needed in a specific situation. Therefore we can speak of induced pluripotent stem cells (iPS cells) being 'reprogrammed' into a state in which they are pluripotent. The cells program the genome, just as the genome programs the cells. This was not intended by the idea of the genetic program.

Changing the DNA sequence can still have severe consequences. It can disrupt the ability of the cell to create the appropriate biomolecules at the right places and at the right time. Hence, the identity crisis of the genome does not lead to a diminished influence of the DNA on development. It only accounts for this influence in rather different ways.

The opacity of the genes

Now we can go back to where we started this chapter: George Church's encouraging call to push forward with personal genomic sequencing. Through his wording ("inspect the genome") and his

---

[38] Gros, 1991.

[39] See also Müller-Wille/Rheinberger 2009.

[40] Turney, 2009, 141.

reasoning he seems positive about the possibility of seeing the genome of a person by sequencing it. Obtaining a genome sequence is no longer so expensive, it is fairly accurate, and it gives us enough useful information – with more than 3000 highly predictive gene tests available, we can all be identified as carriers of some life-threatening risk mutations, and people can learn to cope with this predictive information without it being a major psychosocial risk. This is his argument.

It is interesting to read the online discussion that followed the publication of Church's essay. Several readers, with a diversity of concerns, showed more caution. They pointed to the current limitations to the understanding of genetic information. Understanding is more than just knowing a sequence, they claim. It would mean understanding the consequences for the individual who is sequenced. From the point of view of the individual who wants to know about her or his genes it is not sufficient that genetic information is important *in general* and could do some good for some people *in special circumstances*. It is, for example, unclear what it means for a patient when her physician tells her that her newly detected individual sequence in the 9p21 locus (which is said to be important in assessing the risk for coronary artery disease) is more frequently correlated with people who have coronary artery disease than with others who are healthy. This information might be scientifically correct. But it is difficult, or bluntly impossible, for an individual who is interested in finding out something about herself to read. One commentator found an illuminating metaphor: "The pendulum of interpretation must swing first from 'knowledge' to 'understanding'."[41] The un-aided human brain might be overexposed to a myriad of details of genetic knowledge. In the same way that a photograph overexposed to light does not show clear contours of the objects, providing genetic information with a myriad of genetic details but without a key to make them meaningful make it intransparent to human understanding.[42] For a meaningful

[41] Andras Pellionisz, 16 Oct 2013, at 05:50 pm.

[42] One can see an analogy to Paul Virilio's theory of over-exposition ('Überbelichtung') to media images that make things actually invisible (cf. Virilio 2008).

interpretation the individual would need the aid of specialized computer programs, Pellionisz guesses. Only then could genetic knowledge have an impact on people's health.

But there are other issues that make the genome essentially an opaque, 'intransparent' thing, at least for those who use it, even though more and more single pieces of information or whole sequences become accessible. Barbara Katz Rothman[43] has offered a more sceptical view of genome interpretation, based on fundamental considerations about the interrelatedness of genetic knowledge with human imagination. We take here two of her points which are especially pertinent to our considerations about genetic 'transparency'. The first point has to do with the probabilistic nature of genetic predictions. The usual way of putting this issue is: if we receive the information that we carry a mutation with a high disease risk, we still cannot know whether we will or will not develop the disease. We only know the relative likelihood of our life trajectories: p with the disease and 1-p without the disease. If p is well above minimum and the disease severe (cancer, heart attack, diabetes etc.), we will be well advised to take precautions in order to diminish the likelihood of getting it, or to be ready early enough to fight the disease. But we could also gamble against the odds of our genes. Anyway, the genes do not tell us in which group – the p group, or the (1-p) group – we belong, which is exactly the information that we need.

Rothman suggests a closely related problem, starting from the simple observation that "twins are not the same baby twice."[44] One identical twin can develop type I diabetes, while the other, with the same gene variant on the small arm of chromosome 6, will not have this 'genetic' disease. What is clear in the case of twins – that the two are not the same – is less clear in the case of individual genetic prognoses. Both possibilities – being potentially in the (1-p) group and being potentially in the p group – belong to us. Hence, we imagine being potentially both individuals. But in reality we will become only one of them. Now, both could *potentially* exist, but only one will be real. Two possible cases: while our unreal potential twin

---

43 Rothman, 1998.

44 Rothman, 1998, 26.

will develop the disease, we will not have it. Or the other way round, while the unreal potential twin will remain healthy, we will develop the disease. The one who is different from me still has his or her own genome. As do I. But normally we do not think in terms of imaginary, non-existent twins. The imagination works differently, more simply: if I hear that I have the genetic risk of X% of developing Y, I imagine the whole odds ratio as somehow belonging to me. I say: my genome carries 'a genetic risk' – as if I could be both of them together, my unreal potential twin and myself. The twins are however *different* persons: "they're not having the same experience. *Here* isn't *there*, and nothing is ever the same."[45] The unreal potential twin that we have imagined will be a different person. The nub of the problem is that we confuse the statistical probability with a realistic tendency. We interpret (falsely) the odds ratio as a force that is more or less likely to become effective. We see the gene as a force – which is not what the statistical evidence justifies. When we receive the information that we have a X% genetic risk for Y, there is what we might call an *identity-nonidentity confusion.* As long as we don't know whether we actually have Y, we imagine being a mix of two people, not knowing who we will be: one with Y and one without. The whole concept of a genetic risk, when applied to individual genetic prognosis, carries this difficulty. It is essentially unclear to the imagination, and hence opaque.

And, as Rothman concludes, any decisions will be made with poor information: "Take one cell, one 'blueprint', one 'book' of DNA, and twin it or clone it as you will: you don't get the same person over and over again. Read all the information, decode the book, and you can make predictions, based on knowledge, but you also must acknowledge that you can't know. Decisions will have to be made – about screening and testing and aborting and preventing and treating – always with inadequate information."[46]

A second point is the difference between prediction and explanation: "To diagnose, to predict Down syndrome is not to explain Down syndrome. And to predict the fact of Down syndrome is not to predict the experience of Down syndrome. One foetus so

---

[45] Rothman, 1998, 27.
[46] Rothman, 1998, 29.

diagnosed is not strong enough to survive the pregnancy; another is born with grave physical and mental handicaps and dies very young; and another grows up well and strong and stars in a television show."[47] The same applies to other conditions, including monogenic diseases such as sickle-cell anaemia. Genetics "cannot predict which person will be severely and which mildly affected by the disease."[48] The actual disease is not the predicted disease, and the prediction of the fact of a condition is not to predict the experience of this condition. Genetics has learned to predict (however only per likelihood) but it has limits in explaining what it means, for those actually affected, to have this predicted condition. There are variants in the phenotype and there are variants in the subjective experience. And above all, the objective fact of being affected by a certain condition does not tell us anything about how this condition is experienced. The phenomenology of illness[49] is not the same as the pathology of the disease. Genetic predictions are essentially intransparent both in regard to the actual phenotypic variant and to the experience of the predicted illness.

When we talk about genetic transparency we should therefore ask first of all what should be made transparent. Is it a mutation, a set of mutations, a pattern of mutations? This is actually feasible, but it is more of a challenge to understand what they signify for the life of the person who seeks transparency. Is it the genotype? This is much more difficult, since the genotype is a hypothetical and holistic figuration, which accounts for the real visible 'phenotype' that we are and will be.[50] Is it the essence of us that accounts for who we are and how our body stays alive? Perhaps some people actually look for explanations about themselves as persons, for who they are as embodied beings. And genetic information brings about something similar to Aristotle's explanation of the essence: *to ti en einai* – literally 'the what it *was* to be'[51]. Note the past tense in his formulation ("was"): Aristotle's proposal to explain what we mean by 'essence'

---

[47] Rothman, 1998, 32.
[48] Rothman, 1998, 33.
[49] cf. Carel 2008; Carel/Cooper 2013.
[50] See further remarks on the genotype-phenotype relation below in chapter 2.
[51] Aristotle, Metaphysics VII, 1029b 26.

(*ousia*) is to explain what it means to be "this" from the past time, i.e. from what it "was" to be this. But it is highly speculative to place genetic evidence in this role of the *ousia.*

Is it our lived future? Only in exceptional cases can genes predict with 100% certainty what will happen to us. Or is it genetic risks that we want to see? Yes, that's it. This is what genetics has to offer. But, as we have seen, knowing a genetic risk is not to have clear knowledge about ourselves in the future. Whatever we put into the equation, genetic transparency remains a problematic concept.

Is there a way out of this *aporia* of genetic transparency? We don't think that there is an easy way out, as long as we attempt to reach beyond the sphere of human imagination about the genes to the truth about ourselves. We can however take the imagination more seriously and describe it as such. Genetic transparency is established within the sphere of what we *imagine* genes to be. This is an unsatisfactory path for positivists, but it better describes what is going on in society. And it focuses our attention on the interpretative work people put into the meanings they attach to the genes that we could know about. We have suggested elsewhere to describe this process of interpretation as reflexive embodiment. The reflexively embodied genome is charged with basically two sets of meanings that both differ and interact. "One is the set of meanings that are attached to the genetic in biomedical research and in clinical contexts. For the sake of simplicity, we call this perspective, and what is seen in it, 'genome 1'. The genome, however, is translated and transformed into a related but dramatically different figuration that we call here 'genome 2', which is the genome seen within the lifeworlds of the individuals concerned."[52] Questions are raised and answered within these lifeworlds, such as: 'What does my genetic make-up mean for myself and for my family?' or: 'In what sense "am I my genes"?' Distinguishing between genomes 1 and 2 is a heuristic move that helps to identify these two sets of meanings connected to the genome in socio-cultural contexts. The actual 'real' genome remains the intended object of both, which however can only be grasped within one of these two sets of meanings. People seek genetic transparency within the framework of genome 2, as questions that are meaningful

---

[52] Rehmann-Sutter/Mahr, 2016.

in their lifeworld, and they very often receive answers that are phrased in the sense of genome 1. Translation between genome 1 and genome 2 however is far from trivial.

Every scientist says that the public needs to be educated about the benefits and risks of genome information. But this involves more than education about molecular genetics and biostatistics. As we have seen, the state of the discussion within science is very dynamic: geneticists themselves have difficulty today explaining what the genes are. Such complexity is difficult to communicate, as is the lack of knowledge. Science journalist Jon Turney[53] summarizes the situation like this: metaphors, which would be both scientifically accurate and as revealing like the program or the book of life, are currently lacking. This is a fundamental problem for science communication in the field of genetics. For want of better ones, old metaphors are re-used. The program, the software, the blueprint. Right now we are seeing a broad discussion about 'genome editing'[54]. Editing is a metaphor that suggests DNA is software that contains errors that can be corrected, in the way that a text can be edited. But the re-use of old metaphors also brings out old meanings that lurk behind. As Ina Hellsten[55] has explained, metaphors carry an important temporal dimension. Metaphors are like 'time capsules' (which is itself a metaphor): the 'program' or 'software' images of the genome transport meaning from the genetics of the 1960s. They have deterministic and essentialist undertones and do not do justice to the actual complexity of cellular processes as they are understood today.

And this adds to the lack of transparency, the opacity of the genome. People 'see' an imaginary that is prompted by the metaphors they use; they do not see the genes. The sequences that are revealed, or the mutations that are tested, are written into this imaginative setting and become meaningful in this genome 2 framework. When the metaphors tell an old story, which is no longer scientifically accurate, the genome 2 information value of genetic information can be misleading.

---

[53] Turney, 2009.
[54] Lanphier et al., 2015.
[55] Hellsten, 2009.

The light that is contained in the image of the 'transparency' metaphor leads in precisely the opposite direction. Light, as a metaphor, has always had a connection to truth[56], which is a seductive implication here, where genetics is concerned: It invites us to believe what we are told, not to scrutinize the information value of knowledge about our genes.

Literature

Aristotle's Metaphysics, edited by W.D. Ross. 2 vols. Clarendon Press, Oxford, 1924.

Aristotle, Poetics, edited and translated by St. Halliwell, Loeb Classical Library, Harvard 1995.

Benne, Rob et al.. Major transcript of the frameshifted coxII gene from Trypanosome mitochondria contains four nucleotides that are not encoded in the DNA, Cell 46 (1986), 819-826.

Black, M. Models and Metaphors, Cornell University Press, Ithaca, 1962.

Blumenberg, Hans. Paradigmen zu einer Metaphorologie. In: Rothacker, Erich. Archiv für Begriffsgeschichte Vol. 6, Bouvier, Bonn, 1960, 7-142.

Carel, Havi. Illness, the Cry of the Flesh, Acumen, Stocksfield, 2008.

Carel, Havi, Cooper, Rachel (eds.). Health, Illness, and Disease. Philosophical Essays, Acumen, Durham, 2013.

Cavell, Stanley. The Claim of Reason. Wittgenstein, Scepticism, Morality and Tragedy, Oxford University Press, Oxford, 1979.

Church, George. Improving genome understanding, Nature 502 (2013), 143.

Collins, Francis. The Language of God: A Scientist Presents Evidence for Belief, Free Press, New York, 1996.

Dawkins, Richard. The Selfish Gene, Oxford University Press, Oxford, 1976.

Delbrück, Max. Aristotle-totle-totle. In: Jacques Monod and Ernest Borek (eds.). Of Microbes and Life (Festschrift A. Lwoff), Columbia University Press, New York, 1971, 50-55.

Engels, Eve-Marie. Die Teleologie des Lebendigen, Duncker und Humblot, Berlin, 1982.

Falk, Raphael. Genetic Analysis. A History of Genetic Thinking, Cambridge University Press, Cambridge 2009.

Griffiths, Paul, Neumann-Held, Eva M. The many faces of the gene, Bioscience 49 (1999), 656-662.

Griffiths, Paul, Stotz, Karola. Genes in the Postgenomic Era?, Theoretical Medicine and Bioethics 27(6) (2006), 499-521

Gros, François. Les secrets du gène. Nouvelle édition revue et augmeentée, Odile Jacob, Paris 1991.

---

[56] Blumenberg, 1960

Hellsten, Iina. Metaphors as Time Capsules: Their Uses in the Biosciences and the Media. In: Nerlich, Brigitte, Elliott, Richard, Larson, Brandon (eds.). Communicating Biological Sciences. Ethical and Metaphorical Dimensions, Ashgate, Farnham, 2009, 185-200.

Jacob, François, Monod, Jacques. Genetic Regulatory Mechanisms in the Synthesis of Proteins, Journal of Molecular Biology 3 (1961), 318-356.

Johannsen, Wilhelm. Elemente der exakten Erblichkeitslehre, G. Fischer, Jena, 1909.

Kay, Lily. Who Wrote the Book of Life? A History of the Genetic Code, Stanford University Press, Stanford, 2000.

Keller, Evelyn Fox. The Century of the Gene, Harvard University Press, Cambridge, 2000.

Lakoff, G., Johnson, M.. Metaphors We Live By, Chicago University Press, Chicago, 1980.

Lanphier, Edward et al.. Don't edit the human germ line, Nature 519 (2015), 410-411.

Maasen, Sabine, Weingart, Peter. Metaphors and the Dynamics of Knowledge, Routledge, London, 2000.

Mayr, Ernst. Cause and Effect in Biology, Science 134 (1961), 1501-1506.

Mayr, Ernst. Evolution and the Diversity of Life, Belknap, Cambridge, 1976.

Mayr, Ernst. The Growth of Biological Thought, Belknap, Cambridge, 1982.

Moore, David S.. The Dependent Gene. The Fallacy of 'Nature vs. Nurture', Times Books, New York, 2001.

Moss, Lenny. What Genes Can't Do, MIT Press, Cambridge, 2003.

Müller-Wille, Staffan, Rheinberger, Hans-Jörg. Das Gen im Zeitalter der Postgenomik. Eine wissenschaftshistorische Bestandesaufnahme, Suhrkamp, Frankfurt a. M. 2009.

Nelkin, Dorothy, Lindee, Susan. The DNA Mystique. The Gene as a Cultural Icon, Freeman, New York, 1996.

Neumann-Held, Eva M., Rehmann-Sutter, Christoph (eds.). Genes in Development. Re-reading the Molecular Paradigm, Duke University Press, Durham, 2006.

Oyama, Susan. The Ontogeny of Information, Cambridge University Press, Cambridge, 1985.

Peluffo, Alexandre E. The "Genetic Program". Behind the Genesis of an Influental Metaphor. Genetics 200 (2015), 685-696.

Rehmann-Sutter, Christoph. Was ist ein Lebewesen? Zur philosophischen Herausforderung durch die Molekularbiologie, Scheidewege 23 (1993/94), 142-159 (reprinted in: Rehmann-Sutter, Christoph. Zwischen den Molekülen. Beiträge zur Philosophie der Genetik, Francke, Tübingen (2005), 43-60).

Rehmann-Sutter, Christoph. Instruierte Reproduktion. François Jacobs Konzeptionen des genetischen Programms 1961 bis 1997. Figurationen, Gender Literatur Kultur 4(2) (2003), 29-48.

Rehmann-Sutter, Christoph, Mahr, Dominik. The Lived Genome. In: Whitehead, A., Woods, A., Atkinson, S., Macnaughton, J., Richards, J. (eds.). Edinburgh Companion to the Critical Medical Humanities, Edinburgh University Press, Edinburgh, forthcoming 2016.

Ricoeur, Paul. La métaphore vive, Seuil, Paris, 1975.

Rothman, Barbara Katz. Genetic Maps and Human Imaginations. The Limits of Science in Understanding Who We Are, Norton, New York, 1998.

Schmidt, Kirsten. Was sind Gene nicht? Über die Grenzen des biologischen Essentialismus, Transcript, Bielefeld, 2013.

Shapiro, Robert. The Human Blueprint. The Race to Unlock the Secrets of Our Genetic Script, St. Martin's Press, New York, 1991.

Turney, Jon. Genes, Genomes and What to Make of Them. In: Nerlich, Brigitte, Elliott, Richard, Larson, Brandon (eds.). Communicating Biological Sciences. Ethical and Metaphorical Dimensions, Ashgate, Farnham, 2009, 131-143.

Virilio, Paul. Die Verblendung der Kunst, Passagen, Wien, 2008.

2

Making Genomes Visible

Benedikt Reiz, Jeanette Erdmann, Christoph Rehmann-Sutter

Introduction

In this chapter the book transitions from the metaphorical to the technical. Metaphors will not disappear from sight, since they are abundant: As we saw in chapter 1, 'visibility' and 'transparency', when these words are applied to genomes, are metaphors. In this chapter we will look more closely at the ideas of 'mapping' and 'sequencing'. It will be interesting to ask whether 'mapping genomes' is also a metaphor, or just the description of a certain act of visibilization (a kind of making something visible), the same activity that also occurs in geographical mapping. The genome however is not a landscape. In 1957, C. H. Waddington introduced his famous idea of an 'epigenetic landscape'[1] – explicitly as a metaphor. Geneticists locate bands or mutations on chromosomes, or SNPs, promoters, exons or introns on a DNA sequence.[2] They detect their spatial relatedness and provide a map as an analytical tool, an instrument that creates a certain form of visibility. We can see a map; we cannot see a genome.

In order to describe the processes of visibilization that are involved here, we start by explaining the technological possibilities of creating transparency, as well as some of the risks and factors that affect it. We also indicate associated conceptual, methodological and anthropological questions that will be discussed in later chapters. Visibilization, in the first place, is a technological achievement that uses sequencers, computers, test arrays, algorithms, prints, plots and the like in particular ways and combinations. We begin with the

---

[1] Waddington, 1957.
[2] Gugerli, 2004, 210-218.

current laboratory techniques of visibilization, including high-throughput genomic sequencing, other kinds of omics, and array technology, with a brief summary of the history of these methodologies (2.1). Then, we provide an overview of the possible disease load that individual genomes may carry (2.2), with an emphasis on what is currently known in genetic medicine about monogenic and complex diseases, their correlations with genomic variations, and what trends are expected in the future as knowledge increases.

In the third section (2.3), epistemological questions about visibilization of the genotype will be addressed in more detail. Throughout the history of genetics, the concept of the genotype has evolved and become significantly more complex. There is no direct correlation between a genome sequence and a variation of this sequence (mutation) on the one hand and a phenotype such as a particular disease on the other. Systemic interactions of the entire genome with structural details of the developing organism and its environment occur that account for the information becoming operative as the genotype.

Since the post-genomics era, which began after completion of the Human Genome Project in 2003, a radically new way of investigating genotype-phenotype correlations has been developed through genome-wide association studies (GWAS). In section 2.4, we will explain the principles of this experimental procedure, which produces statistical knowledge about the frequencies of correlations between single nucleotide polymorphisms (SNPs) and a particular disease. However, we will not explain the molecular function of these genomic variations in the cell, as the GWAS approach does not require knowledge of the detailed function of the genome or of the intricate complexities of interactions at the molecular level that are studied in developmental genetics. The statistical SNP-phenotype correlation bypasses or 'bridges' those complexities, steps over the unknown, and produces a new kind of clinically relevant knowledge. If disease risks can be easily tested, new public health approaches based on pre-symptomatic testing and risk-reduction measures will become feasible. With this in hand, the questions can be put more precisely: for whom is the genome made visible in these instances, what kind of visibility is achieved, and what are its limits (2.5)?

The chapter has a dialogical structure, which reflects the science-humanities cooperation that we are undertaking in this book. Sections

2.1, 2.2. and 2.4 have been drafted by a human geneticist (JE), and a biologist (BR); 2.3 and 2.5 by a philosopher of biology/bioethicist (CRS). The text is interspersed with questions that either prompt more explanation or lead to further discussion in the following chapters. These questions are printed right-aligned.

State of the art high-throughput sequencing

DNA, also called the blueprint of life, is a molecule that encodes the genetic information used to build all known living organisms, and is passed from one generation to the next.

You say 'blueprint of life'. What this can mean? Isn't this a description that exceeds what can be said based on experimental evidence? It is an interpretation that sounds a bit metaphysical. As it stands, the sentence echoes the popular science talk that we can read in the newspapers. We also need to understand the term 'build'. You are using it metaphorically here. Organisms are constantly changing; all life comes from life. Hence the building process cannot be like building a house or a machine, in the sense of something being composed from parts and then existing as a composite. In the case of living organisms, it is an ongoing, dynamic process that starts from a living cell – the zygote.

You're right to ask. But let us first explain this further. The information in DNA is stored as a code made up of four chemical bases: adenine (A), thymine (T), cytosine (C) and guanine (G). These bases are like the letters in a book, although with over 3 billion base pairs in the human genome, it would take a few thousand books the size of thick novels to store all of the data; yet the complete genome is present in each and every cell of an organism (except red blood cells).

So every tiny cell has this huge amount of information in it. How can we access this information?

To access this information, different methods of DNA sequencing have been developed over the last several decades, and are used in many different fields such as forensics, anthropology, archaeology, microbiology, biotechnology, epidemiology, and of course, the growing field of human genetics/medicine.

The fact that traits are inherited is common human experience. More specific ideas about heredity were proposed in approximately 500 BC in ancient Greece. Hippocrates believed that acquired characteristics were inherited, and he developed the pangenesis theory, which implies that the whole parental organism participates in heredity. Aristotle thought that blood was the basis for passing characteristics from one generation to another.[3] However, it took almost 2000 years until Gregor Mendel developed a set of rules about the traits of progeny of varietal hybrids in the 1860s that would come to be known as the basis of genetics, by crossbreeding pea plants and observing the resulting phenotypes.[4] A few years later Friedrich Miescher was the first person to isolate phosphate-rich compounds from the nuclei of human white blood cells, which contained DNA and proteins.[5] In the first half of the 20th century, the famous experimental findings by Griffith (1928), Avery, MacLeod, and McCarty (1944), and Hershey and Chase (1952) demonstrated that DNA, and not proteins, carry the genetic information.[6] Only one year later in 1953, Watson and Crick, using crucial evidence from a X-ray crystallography photo by Rosalind Franklin, presented the structure of DNA: a double helix formed by base pairs attached to a sugar-phosphate backbone.[7] Together with Maurice Wilkins, who took the first picture of DNA's molecular structure, these scientists were honoured with the Nobel Price in 1962.

The first attempts to 'read' DNA sequences were made in the early 1970s, and in 1977 two reliable techniques were presented; one by Allen Maxam and Walter Gilbert[8] and one by Frederick Sanger[9]. The Maxam-Gilbert sequencing method uses chemicals to break up DNA in order to determine its sequence. It requires radioactive

---

[3] http://www.britannica.com/EBchecked/topic/228936/genetics.

[4] Mendel, 1865, 3-47.

[5] His et al., 1869, 33-38.

[6] Avery/McLeod/McCarty, 1944, 137-158; Hershey/Chase, 1952, 39-56; Griffith, 1928, 113-159.

[7] Watson/Crick, 1953, 737-738.

[8] Maxam/Gilbert, 1977, 560-564.

[9] Sanger et al., 1977, 687-695.

labelling at one end of the DNA, and purification of the DNA fragment to be sequenced. Chemical treatment then generates breaks at a small proportion of one or two of the four nucleotides, thus generating a series of labelled fragments, from the radiolabelled end to the first cut site in each molecule. The fragments are electrophoresed in denaturing acrylamide gels for size separation, and visualized by autoradiography. The Sanger method (see Fig. 1) is based on the copying mechanism of DNA, which is involved in every cell division, and involves the selective incorporation of chain-terminating dideoxynucleotides by DNA polymerase during DNA replication. The DNA sample is divided into four separate sequencing reactions, containing all four of the standard deoxynucleotides (dATP, dGTP, dCTP and dTTP) and the DNA polymerase; to each reaction, one of the four dideoxynucleotides is added. After several rounds of template DNA extension from the bound primer, the resulting DNA fragments are heat denatured and separated by size using gel electrophoresis, followed by visualization of the DNA bands through autoradiography or UV light. Both of these DNA sequencing methods revolutionized biology, and Sanger, Berg, and Gilbert were awarded the Nobel Prize for their work in 1980.

This breakthrough involved another interesting new idea: the theory that DNA works sequentially and contains information like a text that can be read. The physicist Erwin Schrödinger introduced this idea of a 'code' in 1943.

Can you tell us something about the way DNA encodes information and what it means to 'read' here? What exactly is the difference between sequencing and reading?

The best-studied regions of DNA are the ones encoding for proteins. Here, information is encoded in triplets (codons), meaning that a sequence of three base pairs always codes for a specific amino acid, which are the subunits that make up proteins. First, the respective DNA region is transcribed into messenger RNA (mRNA), an intermediate molecule that is very similar to DNA, and is processed by proteins called ribosomes. For each codon, a specific amino acid is incorporated into an elongating chain, which ultimately makes up a protein. Interestingly, this code is universal for all life forms on earth with only minor exceptions. This is a very important point in biotechnology. 'Reading' is often used to describe the process by which DNA information is used in biological processes, whereas 'sequencing' is simply the technique that transforms the DNA sequence into a human readable format.

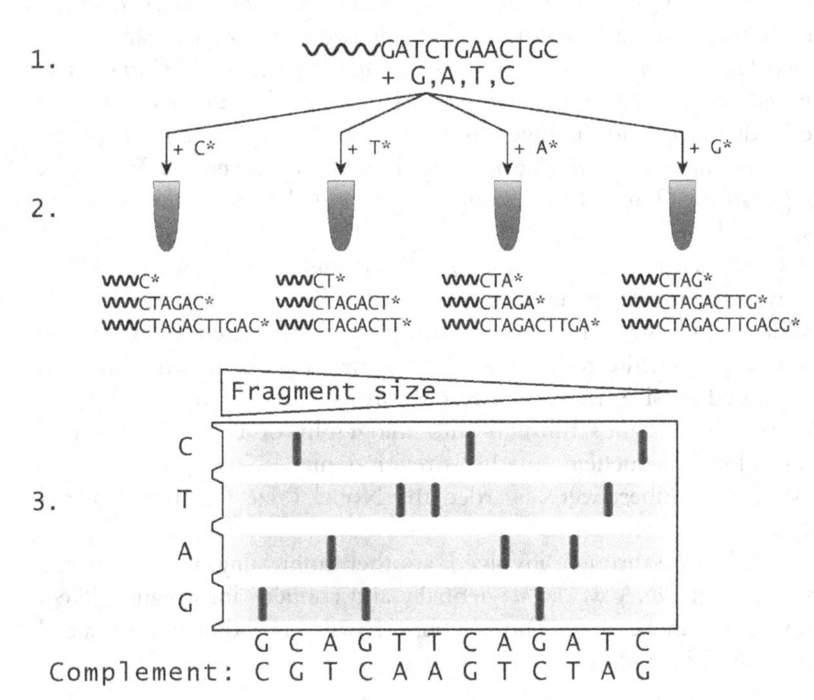

Figure 1: Caption: Sanger sequencing based on chain termination. (1) The four different nucleotides are added to the DNA we want to sequence. (2) Four different reactions are prepared, one for each nucleotide. The corresponding altered nucleotide, indicated by asterisks, is also added. The complementary strand is elongated until an altered nucleotide is incorporated. This happens by chance and results in different-sized fragments.(3) The fragments are separated by size, with the smallest fragments migrating farthest to the right; thus the sequence has to be read from right to left. Due to the complementary nature of the DNA we have to switch A<->T and G<->C to obtain the original sequence.

OK, but then it is 'reading' performed by the cell. So the cell is reading DNA and we are reading DNA as scientists. There must be an interesting relationship between these two types of reading.

The difference is that the cell 'reads' the DNA and can immediately translate the information into proteins and finally into function. We as scientists are now also able to 'read' the DNA, but we cannot translate the information into function easily.

From 1977 we used the so-called Sanger-sequencing method, which allowed a few hundred base pairs to be read, but this required a lot of manual work and the throughput was very low. The Human Genome Project, an international joint project with the goal of sequencing the entire human genome, catalysed the development of cheaper, high-throughput, and more accurate techniques. To this end, in the 1990s, multiple companies developed automated sequencing machines to increase the efficiency of sequencing and to allow multiple sequences to be run in parallel. The Human Genome Project was completed in 2003, with 99% of the human genome decoded. In parallel, a private commercial company, Celera, led by Craig Venter, also presented the sequence of the human genome. They used a faster and cheaper technique for sequencing; however, although they had access to all of the data from the official Human Genome Project, they did not make their own data available in return, causing considerable resentment in the scientific community.[10]

It is interesting that you said 'the' human genome! All scientists spoke of the human genome at that time. Whose genome was it? And is there such a thing as the human genome?

After completing the Human Genome Project, it turned out that there were a lot more differences between genomes than expected. If you look at two individuals, they will differ in about 1 in every 1000 nucleotides. So of course every human individual's genome is different. For the Human Genome Project, a mixture of DNA from anonymous donors was sequenced. What we use today as a reference sequence is the consensus sequence of many individuals worldwide. In this context, a reference must not be mistaken for 'normal' or 'healthy'; rather, it is just the sequence that scientists are referring to when describing an identified variant. Although scientists learned a great deal from these data, expectations were high, and more

---

[10] Butler, 2001, 747-748.

questions were raised than answered. Two years later, in 2005, a new kind of sequencing technique emerged. Next-Generation Sequencing (NGS), more accurately referred to as High-Throughput Sequencing (HTS), changed the whole field of genomics. One major benefit of this technique is the ability to sequence thousands of molecules at the same time. This so-called massive parallel sequencing was made possible by miniaturizing (see Fig. 2). For example, the Ion Torrent semiconductor sequencing technology developed by Life Technologies allows the entire sequencing process to take place on a stamp-sized chip. The total cost of the Human Genome Project was approximately US$2.7 billion[11], and subsequently, the cost of sequencing the human genome has been between $10 and $100 million. However, the cost of new techniques is dramatically dropping to the point that now, in 2015, we are very close to the magic number of $1,000 per genome using Illuminas HiSeq X Ten machines[12] (see Fig. 3). This fall in sequencing prices has made it possible to perform large sequencing projects, including the 1000 Genomes Project[13] in which ~2500 genomes from people worldwide were sequenced, the Exome Aggregation Consortium in which more than 60,000 exomes (the coding regions of the genome) were sequenced[14], and most recently the 2014 British 100,000 Genomes Project[15]. In January 2015, Barack Obama unveiled details of his Precision Medicine Initiative, a bold new research effort to revolutionize how we improve health and treat disease. This project will be launched with a $215 million investment from the US President's 2016 Budget, in the hope that the initiative will pioneer a new model of patient-powered research that promises to accelerate biomedical discoveries and provide clinicians with new tools, knowledge and therapies to personalize medicine and create a treatment approach that takes into account individual variability in genes, environment and lifestyle. Over the last few years, the market for commercial sequencing has grown rapidly and prices have

[11] http://www.genome.gov/11006943.
[12] http://www.genomicsengland.co.uk/the-100000-genomes-project/.
[13] The 1000 Genomes Project Consortium, 2010, 1061-1073.
[14] Exome Aggregation Consortium (ExAC), Cambridge, MA.
[15] http://www.genomicsengland.co.uk/the-100000-genomes-project/.

continued to fall. However, we need to distinguish between commercial sequencing for research and direct-to-consumer services. In 2008, Knome, a company located in Buckeystown, Maryland, United States, also known as the Human Genome Interpretation Company, began offering the first DTC whole-genome sequencing (WGS) service for $350,000 per sample. Since then, the price has dropped to about $5,000 per sample; however, the company recently stopped offering this service. In 2009, Illumina announced personal full genome sequencing service for $48,000 per genome, and as previously mentioned, in 2014 the company launched the $1,000 genome, which refers to full genome sequencing of an individual for $1,000.

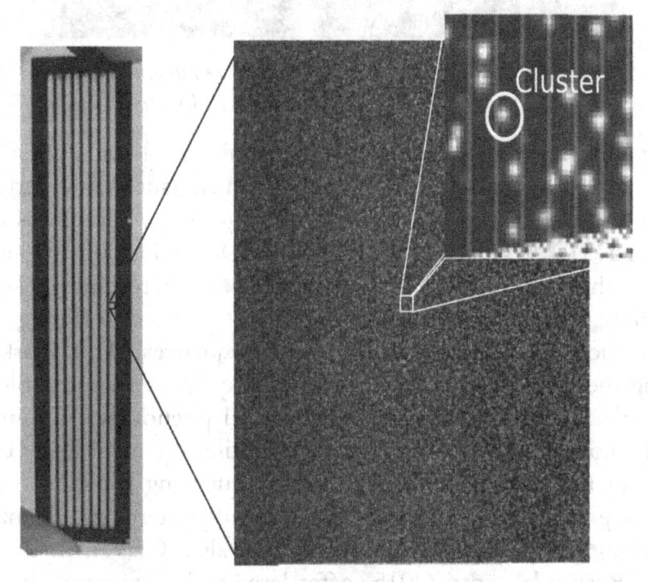

Figure 2: Caption: An Illumina Genome Analyzer flowcell (left) and imaging region or 'tile' (right), with a magnified section showing a cluster. Each spot represents one cluster from a single DNA molecule being sequenced.[16]

---

[16] Whiteford/Skelly/Curtis et al., 2009, 2194-2199.

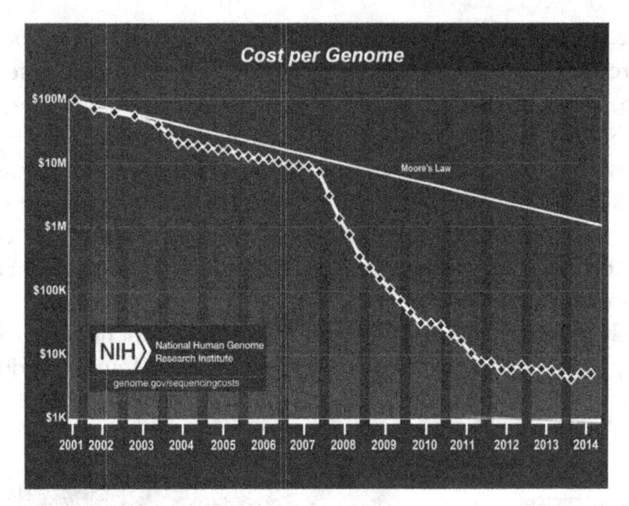

Figure 3: Caption: Cost per sequencing of one entire human genome. (Source: National Human Genome Research Institute)

Currently, more than two dozen companies worldwide offer DTC personal sequencing services, some of which (such as Interleukin Genetics or Genelex) target a very specific group, while others (such as DNA DTC) have a broader portfolio. Depending on the service offered the prices differ substantially. Interestingly, several companies, such as Diploid, have just started offering data interpretation of existing whole-exome sequences. This works by allowing the individual to upload exome sequences as a text file and to provide information about the observed phenotype. Within 1–3 days, the company delivers a report suggesting the most likely causal variant for the phenotype. Several large sequencing facilities, such as the Beijing Genomics Institute (BGI) headquartered in Shenzhen, Guangdong province, China, and the Australian Garvan Institute of Medical Research, today (2015) offer large-scale (minimum of 1000 samples) WGS services for a reasonable price of about $1,500 per sample and a turnaround time of about 4–8 weeks.

Of course, the current methods are not the end of the line, and in fact the next generation of sequencing techniques is close to being unveiled and promises even lower costs, better quality, and faster turnaround times. The most promising candidate is the pocket-sized MinION from Oxford Nanopore Technologies, which was first proposed in 1990. Instead of amplifying the target DNA first, this method allows the sequencing of very long single molecules. Briefly, a single DNA strand is passed through a nanopore, which is smaller than 1 nanometre in diameter. This induces a conformation change in the nanopore, which leads to a change in electric current that can be monitored. These changes in current are very specific to the state of the nanopore, and thus for the nucleotide that is passed through it. This technique will help to overcome many of the difficulties with the current methods (e.g. no amplification of DNA fragments is needed before sequencing). Although the initial results are now being published, this methodology is currently available only to selected researchers for testing purposes.

These new sequencing methods have given us insight into the sequence and function of our genome, although there is still a great deal more to discover. Of course, all these discoveries are not interesting from a purely academic point of view alone; they have also given rise to the field of medical genetics.

What is known about the genome's disease load and the medical implications of human genomics?

Despite the fact that genetics and genetic testing have played a role in human medicine since the mid-20th century, it was not until the development of automated sequencers that they became an essential part of modern medicine, which was reinforced after the development of high-throughput techniques. The actual method used always depends on what one wants to know. For some diseases, known genes affect the condition, so either a single gene or a specific set of genes is sequenced using so-called panel sequencing.[17] These panels are designed to screen a few to a hundred genes, depending on what one wants to know. The advantage of this technique is that the whole genome/exome does not have to be sequenced, which decreases the price and workload. Other situations, for example if no initial information is available, might demand sequencing of the whole genome or exome. Currently, whole-exome sequencing (WES), which includes the portion of DNA used by cells to produce proteins

[17] Mauer/Pirzadeh/Robinson/Euhus, 2013, 407-412.

(exons), is used more widely than WGS for two main reasons. First, WES needs only the RNA-coding region, which is about 1% of the total genome, making exome sequencing much cheaper. However, the emergence of new techniques might soon change this, making WGS the preferred method (see 2.1). Second, exonic sequences are the best studied and understood as they contain the information used to make proteins. There is still much to learn about the function of the rest of the genome, and even if genetic variants are found, it is hard to derive a function and to come up with the appropriate clinical treatment. Until a few years ago, people even believed that most non-exonic regions in DNA sequences might not be functional at all, and as such, they were often referred to as 'junk DNA'. Nevertheless, projects like ENCODE[18] have tried to determine the role of the DNA regions that are not transcribed. It is currently thought that these regions play a key role in the genetic machinery, but their precise functions are poorly understood.

So not only are those approximately 22,000 genes important for life, health and disease, but the other parts in between the genes are important as well, but in different ways than via protein synthesis?

Despite the fact that it is still not quite clear how many genes there really are, and what exactly falls under the definition of gene, you are right. As previously mentioned, each cell contains the same genetic code, but obviously a liver cell differs from a heart cell. The key to this differentiation lies in the expression of different genes. Some genes are cell-specific and are only expressed (e.g. transcribed into a protein) in specific cells. This is regulated by regions in the genome called enhancers, silencers, promoters, and many more. There is a very complex network of regulatory DNA elements, although we are far from understanding the big picture.

To distinguish disease-causing alterations from the bulk of neutral changes in the genome, we need something with which to compare the derived sequence. For this purpose, reference sequences are available from healthy controls worldwide, which is important because some alterations are population-specific. These non-

[18] The ENCODE Project Consortium, 2012, 57-74.

malignant variants originated in different regions through human evolution; it is therefore important to take into account the genetic background of the patient.

There are typical Japanese genomes, say, or Northern African ones?

This is not an easy question to answer, because individuals from different populations can show less difference in their genomes than two individuals from the same population,[19] a fact we need to keep in mind when talking about races. But there are still population-specific variants, individual changes with respect to a universal reference genome, which are common in one population and absent or very rare in others. An important concept for medical genetics is that these variants can interact with disease-specific variants. For example, let us say there is a population-specific variant 'A'. Now a novel mutation occurs resulting in variant 'B'. This variant may cause disease only when paired with variant 'A', making individuals who carry 'A' more prone to the effect of this variant. In many cases, it is even better to take the sequence information of family members into account, either in large pedigrees with affected and non-affected members, or alternatively on a very small scale in which only one child and his or her parents are analysed. In this way, family-specific variants can be ruled out. For example, if a child is born with a severe phenotype due to a genetic variant, diagnosis would be difficult if the child's sequence were compared to the reference sequence only. As we saw before, any two individuals differ in 1 in 1,000 nucleotides. So we would expect to find about 3 million variants in the child, but which is the one variant we are looking for? If the parents are sequenced as well, their variants can be subtracted from the ones found in their child, leaving only about 60 variants that have occurred de novo. If the parents are healthy and the phenotype is not the result of a combination of the two parents' variants, we would expect 1 of these 60 variants to be causal.

There is a broad spectrum of clinical sequencing applications that complement existing medicine. Perhaps the most important is the use of WES or WGS to obtain a final diagnosis for a patient, very often

---

[19] Witherspoon/Wooding/Rodgers et al., 2007, 351-359.

one with a rare disease. For more than 2,500 of the several thousands of rare diseases (each affecting fewer than 1–5 per 10,000 individuals), a molecular genetic test is on the market; however, millions of people worldwide live without a specific diagnosis for their rare disease. The Online Inheritance of Man (OMIM) database, a catalogue of human genetic diseases, reports about 1,500 Mendelian genetic disorders, albeit with no causative gene mutation. However, as WGS and WES become increasingly established as routine methods in the clinic, and as the bioinformatics pipelines become optimized for data analysis, the number of causative genes that can explain a patient's phenotype will grow rapidly.

Testing for genetic defects as a method of prevention[20] is also important. A common example is newborn screening or even prenatal diagnosis to identify genetic disorders that can and should be treated early in life, often even before they occur. However, prevention is not limited to newborns. For example, individuals with a family history of a genetic disorder can be treated preventively if it is known that they carry a genetic defect, before any clinical symptoms are apparent. In addition, screening for common genetic disorders such as some types of cancer (e.g. breast cancer) is well established. These findings can help physicians select the optimal care for these patients. In cases of metabolic disorders or an increased risk for events such as myocardial infarction, changes in lifestyle – especially in diet and exercise – could be sufficient. In more severe cases, medical therapy is also supported by genetic testing. Even if two patients present with the same clinical phenotype, they might need different medications if genetic analysis indicates different underlying molecular mechanisms. Additionally, the dosage of medication can be adjusted based on the knowledge obtained from testing specific genes. The variation in genes involved in drug metabolism is the basis of pharmacogenetics, the study of genetic variations that influence an individual's response to drugs. This can protect the patient from long and often painful trial and error adjustments, especially in cancer treatment.[21]

---

[20] Collins, 2010, 674-675.
[21] Verma, 2012, 1-14.

For many patients, for instance, it would be beneficial to know their *cytochrome P450 (CYP)* genotype in detail (humans have 57 *CYP* genes), because some of its variants can markedly influence drug metabolism, while others can lead to unpleasant side effects and drug interactions.[22] The CYP proteins are involved in the degradation of toxic compounds in the liver. For example, if one drug inhibits the *CYP*-mediated metabolism of another, the second drug may accumulate within the body to toxic levels. In such situations, drug interactions may necessitate dosage adjustments or choosing drugs that do not interact with the *CYP* system. In order to do this, however, knowledge of the *CYP* genes is a prerequisite. This is just another example demonstrating that knowledge about certain genomic details can be in a potential patient's or an actual patient's best interests.

The epistemology of 'visibility'

What it means to 'know' something via tests, maps or sequences raises some philosophical questions. What is this knowledge about? Is knowledge of a sequence knowledge about the genome, or knowledge about our bodies? These two things are not the same. As patients or potential patients we are interested in knowledge about our bodies. If all we learn is something about the genome it would be less relevant, at least if it has no implications for the body. However, if knowledge about bodies can be provided through genomic knowledge, does that tell us only something about the present state of the body, or also something about its future potential and development?

In genetic testing or sequencing of an individual, a previously invisible aspect of the body becomes visible. The fact is established that an individual's DNA, which is contained in the chromosomes in the nuclei of his or her cells, has a particular characteristic at a particular locus within the sequence. This is what is actually tested – a mutation, a translocation, a sequence pattern. However, this is not what the individual and the genetic analysts are ultimately seeking;

[22] Bhattacharyya/Sinha/Sil, 2014, 719-742.

rather, they are looking for the gene or the genotype. It is the genotype, not the phenotype, that provides predictive knowledge, because it is the genotype that determines the hereditary potential and limitations of the individual. Thus, this mutation, translocation and sequence pattern represent a genotypical feature of this individual.

> What, then, is the difference between the DNA and this strange theoretical figuration of a genotype? And what kind of knowledge is offered when the DNA and genotype are made accessible?

We speak here of visibilization in the rather general sense of making the genotype accessible to our senses. Parts of it might become literally visible as a pattern of coloured spots on a computer screen or as a plot on a sheet of paper (see figure 2), but we do not want to overemphasise the importance of the visual sense. Here, visibility provides a kind of knowledge; however, the question is what this kind of knowledge is. (The genotype could also be made *audible*, although similar questions about the epistemology of audibilization would then need to be raised.) For instance, if a newborn whose parents are both carriers of a cystic fibrosis mutation tests positively for the F508del mutation in the gene that encodes the Cystic Fibrosis Transmembrane Conductance Regulator (CFTR) protein on both of its chromosomes, it is not only known that the child's CFTR gene actually carries this F508del mutation on both of its chromosomes – this is ultimately a phenotypic characteristic of its DNA – but that it will very likely develop the severe symptoms of this disease later in life. The mutation is a causal determinant for the future development of the child's body, and this causal potentiality makes it a part of the genotype. In this case of a homozygous mutation in the CFTR gene, it is certain that the child will actually develop the disease. Early therapy could alleviate and postpone the symptoms. In other cases such as cancer-related mutations in the BRCA genes, certain detectable mutations only provide a statistical measure of a certain likelihood that a disease might develop, which, as is widely known since Angelina Jolie's coming out in the *New York Times*,[23] can also be

[23] Jolie, 2013; Jolie Pitt, 2015.

a basis for deciding to have preventive operations such as mastectomy or ovarectomy.

The distinction between phenotype and genotype is not without theoretical difficulties, which also affect the understanding of the epistemic significance of genetic findings for the present and future life of an individual. These difficulties reach beyond the notorious difficulty of interpreting statistical probabilities for an individual case: if an individual has a genetic risk p for a certain condition, the individual knows that he or she might belong either to a group of the (1-p) x 100% unaffected individuals or to a group of the p x 100% affected ones, but the probability does not tell us to *which* group he or she actually belongs. Instead, the difficulties with the phenotype-genotype distinction with which we are concerned here are conceptual, and are related to the history of the concept of the genotype.

As mentioned in chapter 1, it was the Danish botanist Wilhelm Johannsen who introduced the term genotype in 1905, a few years before he also coined the related term 'gene' in 1909. As Peirson (2012) pointed out, Johannsen criticized what he called the 'transmission conception of heredity', which had prominent supporters in his time. This was the idea that the characteristics of individual organisms are transmitted directly to their offspring. In contrast, Johannsen's concept of genotype was ahistorical; he postulated that the phenotypic characteristics of a mother and her offspring would arise under the influence of the same hereditary disposition, passed from generation to generation, and would be immune to the environmental circumstances in which they were expressed. According to Johannsen, the genotype of an organism gives rise to the organism's phenotype through the process of development, under the influence of the environment.[24] The introduction of the genotype-phenotype distinction was a fundamental step in the successful development of genetics, and later, molecular biology. It is important to keep in mind that Johannsen's pure-line breeding experiments with barley and bean involved Mendelian traits and factors. These are the explanatory references that appear on the two sides of a hereditary table, and which explain

---

[24] Peirson, 2012; Falk, 2009.

the frequency of, say, the red and white colours of a flower. These factors are provided by the previous generation through the gametes.[25] Genotype is a term that encompasses all of these inherited factors. Therefore, it is clear that the genotype must be ahistorical, as Peirson notes, since it remains constant through the life of the individual, and may even remain the same over generations, giving rise to all of the individual variants and concrete shapes that individuals develop in their interactions with the environment. For humans, the genotype would be the same for all of our hypothetical identical twins (or a hypothetical 'clone' of ourselves), despite the fact that each of these individuals would have their own unmistakable identity and a very similar, but still different body. Therefore, the genotype is not just a causal explanation of the individual, but represents a type, in contrast to a token.[26] The word 'geno-*type*' is well chosen.

But in what sense is the genotype a type and the individual a token? We have just seen one explanation: the type is the set of all inherited factors that remain constant throughout individual development and that could also give rise to another individual. The relationship between token and type, however, could also be explained as a relationship between all possible phenotypes of an individual, that vary depending on its state of development. Goodenough[27] suggested that the term be used in this sense: "The genotype of a female mammal, for example, includes a number of genes that do not find expression in the phenotype until sexual maturity." (p. 87). In other words, the phenotype refers to a stage in developmental time, and an individual goes through many phenotypes during his or her lifetime. Harvard biologist Richard Lewontin, following a current custom to invent '-ome' words to express a totality, introduced the term 'phenome' to denominate all shapes of the individual during all developmental stages from embryo to adulthood to death, and took the whole phenome as the token. The phenome is in turn related to the genotype as a descriptor of "the set of physical DNA molecules inherited from the organism's

---

[25] Johannsen, 1911, 129-159.
[26] Lewontin, 2011.
[27] Goodenough, 1978.

parents" (Lewontin 2011). Lewontin's description of genotype refers to a type that encompasses the whole set of phenotypes during a lifetime (the 'phenome'), whereas Goodenough's description of genotype refers to a type that encompasses the developmental stage as a sum of all traits that are actualized in this stage.

These conceptual issues are blurred in popular contemporary definitions of genotype, which explains it simply as "the genetic constitution of an organism" (Encyclopaedia Britannica 2015; the same words are also used in the glossary of an early textbook by James Watson 1976, p. 703). The genetic constitution refers to an underlying causality that together with environmental factors (and coincidence) explains the actual development of an individual. Therefore, the genotype may actually represent very specific individual hereditary potentials and limitations, which are anything but typical. Today, many would say that the individual DNA contains the genetic make-up of the individual, and they would not care too much about these rather philosophical token/type questions. Genetic tests or sequencing just reveal particular details about *this DNA*.

However, the questions cannot be silenced. The particular details of the DNA tell us something about the potentials and limitations of the future development of an individual. It is still the question of the genotype that motivates genetic testing. It is not our aspiration to learn more about hidden aspects of our phenotypes, provided they do not tell us anything about the potential and limitations of future development. Somebody who gets her or his genes tested wants to know what could or will happen. In diagnostic tests, we hope to obtain information about the cause behind an observed phenotype, i.e. what potentials have contributed to its realization. The epistemic value of genetic information is to reveal causes behind a phenotype, explanations and differentiations between similar phenotypes (for instance, similar disease symptoms resulting from different causes), or potential and limitations for future development. This is all knowledge about possibilities, not about realities, about underlying causes, not about their effects. Hence, genetic knowledge always works on a constructed level of explanations, in the *modus potentialis*; it illuminates and describes reality in a particular new way, and makes the corpo-reality of an individual 'transparent' so that an underlying level of causes shines through it. Thus, corporeality incorporates another level of reality. It has an ontologically inherent level of potential that can explain its development.

The visibility of an individual's genetic features (made visible via probing, testing, mapping, sequencing), therefore, alters how we look at the bodies of individuals, ourselves and others. It characterizes the distinct *genetic gaze* on the world of living things. What is actually visible (the manifest physical properties, Lewontin's 'phenome') is always the effect of an underlying genetic process, in which the genome is involved. However, the genome itself, in the sense of the DNA molecules, is also a part of the physical body. But it is not in that function, as a physical component of cells, that it has become an object of genetic investigation and visibilization. What we are looking for is the genome as a source of genotype information that can be derived from the body. This is perhaps an explanation of why genetic information always appears to us as a bit ethereal in some way. It does not appear as a description of a positive reality. It does not tell us what we see. It comes as a (partial) explanation of this visible reality, or as a prediction of what could become corporeal in the future.

This is in stark contrast to a map. In the history of twentieth-century genetics, linkage maps, chromosomal maps and maps of genomes have played a key role.[28] A map does not provide causal explanations but gives a spatial orientation. In one sense, any map is a representation of something. A city map, for instance, represents the streets, squares, metro lines etc. that are built in the city. But it does not give access to the city's real complexity. In order to be useful it must be selective, and sometimes it must even depict some things at a higher magnification than others. Sergio Sismondo[29] compares it with the map of the London Underground, which magnifies distances at the core and shrinks those in the periphery. Geometrical fidelity is abandoned for utility. Similarly, a map of the human genome represents it in a way, to whatever extent selectively. It is useful as an instrument that provides orientation. But both in its representative and in its instrumental functions a map provides information. One map that gained particular significance in the development of present-day genomics was the HapMap. This is a collection of the common patterns of individual sequence variations in the genomes of

---

[28] Gaudilliere/Rheinberger, 2004a; Rheinberger/Gaudillière, 2004b.

[29] Sismondo, 2004, 203-209.

the present human population on Earth and thus a key resource for researchers. It takes as a reference not the diploid genomes (with two of each chromosomes in each cell) but the single, haploid chromosomes.[30]

GWAS as an approach to establish correlations

We have learned a lot from the 1000 Genomes Project[31] about the variability of the human genome. In 2012, in the final phase of the 1000 Genomes Project, the consortium reported the genomes of 1,092 individuals from 14 different populations. The genomes were constructed based on low-coverage whole genome and exome sequencing data. In total, the consortium reported 38 million single nucleotide polymorphisms (SNPs – pronounced "snips"), 1.4 million short insertions and deletions ("indels"), and more than 14,000 larger deletions or insertions. This is currently the most comprehensive catalogue of human genetic variability available, and captures up to 98% of all SNPs with a frequency of at least 1%.

You are talking about a catalogue of human genetic variability. What does this mean? Can anyone look up this genetic variability in a database? Does it provide personalized information about the participants of the 1000 Genomes Project? Is this public knowledge?

The variants identified by the Human Genome Project and the 1000 Genomes Project are indeed publicly available. At www.1000genomes.org you can download the latest version of the human reference genomes, and at www.ncbi.nlm.nih.gov/SNP/ you can find detailed information on every single SNP reported to date, including information about the frequency of the variant in different populations, its exact location on the human chromosome, and information about its potential pathogenic or functional relevance. However, you cannot find personalized information about the participants of this project. Knowledge of the genetic variability on a

---

[30] International HapMap Consortium, 2003, 789-796.
[31] 1000 Genomes Project Consortium, et al. 2012.

population-wide basis enables common and low-frequency variants to be analysed in individuals from diverse populations, including admixed ones. It is this comprehensive catalogue of human genetic variability and its precursors established by the Human Genome Project, as well as the high-throughput technologies developed to assay millions of SNPs simultaneously, accurately, and rapidly in thousands of individuals, patients and healthy people, that has allowed the design of hypothesis-free GWAS, which since 2005 has revolutionized the genetics of common, complex diseases.

What do you mean by 'hypothesis-free' GWAS? Research normally works with hypotheses that are then tested.

For more than 20 years, researchers conducted so-called candidate gene studies, meaning that based on assumptions about the functional role of a specific gene with regard to a specific disorder, single genetic variants were genotyped in samples of cases and controls. At that time, the sample size was small compared to the sample sizes that are currently analysed. From the current point of view, all of these candidate gene studies were heavily underpowered, and it is not at all surprising that almost none of the results from these reported association studies hold true today. However, there are more than 5,000 papers in the literature describing candidate gene studies for a broad range of disorders including coronary artery disease (CAD), hypertension and diabetes. All of these candidate gene studies were driven by a specific hypothesis based on limited knowledge about the underlying pathophysiology. A GWAS is not driven by a specific hypothesis, and that is essential. In contrast, the complete genome is studied without any assumption about the involvement of a specific gene in the pathophysiology of the phenotype of interest. A GWAS is a method for the discovery of genetic links that you cannot know of beforehand.

I can understand the difference between the candidate gene approach and GWAS; however, could you explain in more technical detail how such a genome-wide analysis is actually performed?

First, the case and control samples must be defined. Over the last 10 years, we have learned that both samples should be as homogenous as possible, meaning that the phenotype of the patients needs to be defined as comprehensively as it can be to avoid too much heterogeneity, and the control sample should be age- and sex-matched. In addition, it is important to avoid population stratification, meaning that the case and control samples should be of the same ethnicity to decrease heterogeneity due to population admixture. Second, the samples must be genotyped with one of the commercially available arrays from companies such as Illumina or Affymetrix. For example, Illumina's genotyping technology, "BeadArray microarray technology", is based on silica beads that are covered with hundreds of thousands of copies of a specific oligonucleotide that act as the capture sequences for a given assay. This technology allows the simultaneous and parallel genotyping of hundreds of thousands of SNPs in thousands of individuals. After genotyping, the statisticians examine the frequency of every single SNP, and compare it between the case and control sample. If one type of variant is more frequent in people with the disease, the SNP is said to be associated with the disease. Typically, a single correlation is considered significant when the p-value is ≤ 0.05.

This sounds rather arbitrary to me. Could you please explain this in a little more detail?

To start with: as a scientist it is almost impossible to prove anything is true. All we do is show that things are false. In other words, instead of proving any of our hypotheses, we try to prove that all of the alternatives are wrong. In statistics you call that the 'null hypothesis'. The p-value is the evidence against the "null hypothesis". The p-value itself does not tell you that the "null hypothesis" is correct or right, but only if there is significant evidence to reject it or not. A p-value under 0.05 is commonly considered significant if you test only one "null hypothesis". However, in GWAS, the number of tests is of the order of 10^6 (number of tests = number of independent SNPs), meaning that multiple testing correction is needed. The most conservative correction is the Bonferroni correction for multiple testing; therefore, as the gold standard in GWAS, only a p-value $\leq 5 \times 10^{-8}$ is considered genome-wide significant. In general, the study design for GWAS is simple and works best with common SNPs and common complex diseases such as coronary artery disease, diabetes and hypertension. This is why GWAS has been described as a "revolutionary approach".

This revolution can easily be confirmed by simply counting the number of publications that have reported results from GWAS since 2005. Only a handful of GWAS reports were published in 2005. Since then, the numbers have dramatically increased each year, with more than 1,500 GWAS reported annually. It is not only the number of GWAS that have increased over the last 10 years, but the number of genome-wide significant SNPs for a specific trait reported in each publication. In total, the GWAS catalogue, hosted at the National Institutes of Health, reports 15,396 SNPs showing genome-wide significant associations with any trait (March 2015).

When we know that thousands of SNPs genome-wide are significantly associated with a disease, what is the actual benefit for the patient?

In general, GWAS have greatly advanced our understanding of the genetics of complex diseases. Take coronary artery disease (CAD) as an example: we now know that more than 45 common SNPs genome-wide are significantly associated with this life-threatening disease.[32] Besides these common SNPs we have identified several rare variants, mainly in extended families, that lead to premature CAD.[33] Only one third of these variants are also associated with traditional risk factors such as high levels of lipids or hypertension. This means that two thirds of the variants may act through unknown pathomechanisms, thereby paving the way for completely new therapeutic options. There are already examples in the literature in which the identification of SNPs in genomic regions, which indicate potential disease-related genes, has led to the identification of new therapeutic targets.[34] Overall, we have learned a great deal about the biology of diseases through GWAS, but we must admit that for more than 90% of all reported SNPs, the underlying pathomechanism is still unclear. However, we should not underestimate the value of the information obtained from these GWAS, because all over the world, scientists are now working on unravelling the pathomechanisms that

---

[32] Kessler et al., 2013.

[33] Erdmann et al. 2013.

[34] Manolio et al., 2013, 258-267.

underlie these associations, and it is undeniable that valuable insights into disease biology will be gained, which will hopefully be beneficial to patients.

Discussion

In addition to the novel insights into disease biology, GWAS data give the patient information about the individual risk for a disease. However, individual disease-associated SNPs confer only modest disease risk, although individual combinations of several risk SNPs (genetic risk scores) might increase the risk substantially. Using genetic risk scores and profiling individuals in highest and lowest relative risk groups might lead to the development of population-based risk screening and stratification programs. Today, companies such as 23andMe, DeCODEme and Navigenics offer consumers the opportunity to access their own genetic data by genotyping genome-wide arrays, and provide personalized interpretations of their data based on genetic associations reported in the literature. However, for almost all diseases, further studies are needed to validate current risk prediction models. There is substantial concern regarding the practicality of using genetic risk scores in the context of risk prediction. Many of these concerns focus on the poor predictive value of currently known markers when used in SNP-based risk prediction models, or their limited incremental value when used in conjunction with non-genetic risk factors for disease. In 2003 the American Food and Drug Administration stopped 23andMe from selling their health-related tests to consumers in the United States. DeCODEme and Navigenics also stopped selling their tests after concerns about test validity[35]; but 23andMe is currently marketing its health-related tests in other countries[36].

The debate about direct-to-consumer marketing of genomic tests (see chapter 4) shows that societies need to clarify for whom the genomes should be made visible in these instances. After all, genomic information is personal information, and who else apart from the

---

[35] Cussins, 2015.
[36] Gibbs, 2014.

person concerned can claim a right to hinder her or him from investigating this information? There are delicate questions to be resolved around the interpretation of the right to know, which implies a right to receive *accurate* information and not to be misled or lured into a contract with false promises.

We need to discuss and further clarify the question of what kind of visibility can actually be achieved by genetic sequencing, mapping and testing. What is seen in the genes is strongly dependent on the expectations and assumptions of those who order a test, read the map, or, as we have sometimes said in this chapter, 'sequence somebody'.

A genetic sequence provides information about the structure of a DNA molecule in one cell of an organism. When it is written as a series of letters A, C, G and T, the sequence provides visibility of that stretch of DNA that has been sequenced. It can be accurate (true), and it can sometimes also contain errors due to mistakes that may occur in the sequencing process. From the point of view of the individuals sequenced, in particular (potential) patients, however, another kind of visibility counts: the visibility of their genotype. They may see the genomic sequence as a sign that tells them about their future, i.e. about the potential and limitations of their bodies with regard to their future life.

Genomic information in this context has some obvious limitations: (i) Genotypical information, even if accurate, is probabilistic information, i.e. it does not tell us whether somebody will develop a disease and when, but only that he or she is more or less *likely* to do so, compared to the population average. (ii) The meaning of genotypical information for the future prospects of the sequenced individual's life depends on meaningful interpretations. The sequence itself does not yet have a fixed meaning in the lifeworld. (iii) Genetic factors interact with other genetic factors, with epigenetic factors, with environmental influences, with the past and future clinical history, and more. Genetic factors alone will rarely lead to a deterministic prognosis. (iv) The information about one's own DNA that becomes available is typically incomplete and partial. For instance, we know only about the SNPs that are tested, not all that actually exist in our genomes.

Literature

1000 Genomes Project Consortium, Abecasis GR, Auton A, Brooks LD, DePristo MA, Durbin RM, Handsaker RE, Kang HM, Marth GT, McVean GA. An integrated map of genetic variation from 1,092 human genomes. Nature. 491 (2012), 56–65.

Avery, O.T., MacLeod, C.M., McCarty, M. Studies of the chemical nature of the substance inducing transformation of pneumococcal types. Induction of transformation by a deoxyribonucleic acid fraction isolated from pneumococcus type III, Journal of Experimental Medicine 79 (1944), 137–158.

Bhattacharyya, S., Sinha, K., Sil, P.C. Cytochrome P450s: mechanisms and biological implications in drug metabolism and its interaction with oxidative stress, Current Drug Metabolism 15(7) (2014), 719–742.

Butler, D. Publication of human genomes sparks fresh sequence debate, Nature 409(6822) (2001), 747–748.

Collins, F. Has the revolution arrived? Nature 464(7289) (2010), 674–675.

Cussins, Jessica. Direct-to-consumer genetic tests should come with a health warning, The Pharmaceutical Journal 294(7845) (2015). URI: 20067564.

Erdmann et al.. Dysfunctional citric oxide signaling increases risk of myocardial infarction, Nature 504(7480) (2013), 432-436. doi: 10.1038/nature12722.

Exome Aggregation Consortium (ExAC), Cambridge, MA (URL: http://exac. broad institute.org).

Falk, Raphael. Genetic Analysis. A History of Genetic Thinking, Cambridge University Press, Cambridge, 2009.

Gaudillière, Jean-Paul, Rheinberger, Hans-Jörg (eds). From Molecular Genetics to Genomics. The Mapping Cultures of Twentieth-Century Genetics, Routledge, London, 2004.

'genetics'. Encyclopædia Britannica Online. Encyclopædia Britannica Inc. (URL: http: //www.britannica.com/EBchecked/topic/228936/genetics). Accessed 3 March 2015.

Gibbs, Samuel. DNA-screening test 23andMe launches in UK after US ban, The Guardian, 2 December 2014.

Goodenough, Ursula. Genetics. 2nd ed., Holt, Rinehart and Winston, New York, 1978.

Griffith, F. The Significance of Pneumococcal Types, The Journal of Hygiene 27(2) (1928), 113–159.

Gugerli, David: Mapping. A communicative strategy. In: Jean-Paul Gaudillière and Hans-Jörg Rheinberger (eds). From Molecular Genetics to Genomics. The Mapping Cultures of Twentieth-Century Genetics, Routledge, London, 2004, 210–218.

Hershey, A.D., Chase, M. Independent functions of viral proteins and nucleic acid in growth of bacteriophage, Journal of General Physiology 36 (1952), 39–56.

International HapMap Consortium. The International HapMap Project, Nature 426 (2003), 789–796.

Johannsen, Wilhelm. The Genotype Conception of Heredity, The American Naturalist 45(531) (1911), 129–159.

Jolie, Angelina. My Medical Choice, The New York Times, 14 May 2013.

Jolie Pitt, Angelina: Diary of a Surgery, The New York Times, 24 March 2015.

Kessler et al.. Genetics of coronary artery disease and myocardial infarction, Current Cardiology Reports 15(6) (2013). doi: 10.1007/s11886-013-0368-0.

Letter I; to Wilhelm His; Tübingen, February 26th, 1869. In: His, W. et al. (eds). Die Histochemischen und Physiologischen Arbeiten von Friedrich Miescher—Aus dem wissenschaftlichen Briefwechsel von F. Miescher, Vol. 1, F.C.W. Vogel, Leipzig, 1869, 33–38.

Lewontin, Richard. The Genotype/Phenotype Distinction, The Stanford Encyclopedia of Philosophy (Summer 2011 Edition) (URL: http://plato.stanford.edu/archives/sum2011/entries/genotype-phenotype/). Accessed 10 March 2015.

Mauer, C.B., Pirzadeh-Miller, S.M., Robinson, L.D., Euhus, D.M. The integration of next-generation sequencing panels in the clinical cancer genetics practice: an institutional experience, Genetics in Medicine 16(5) (2013), 407–412.

Manolio et al., Implementing genomic medicine in the clinic: the future is here, Genetics in Medicine 15(4) (2013), 258–267.

Maxam, A., Gilbert, W. A new method of sequencing DNA, Proceedings of the National Academy of Sciences U.S.A. 74 (1977), 560–564.

Mendel, Gregor. 1866. Versuche über Pflanzenhybriden, Verhandlungen des naturforschenden Vereines in Brünn 4 (1865), 3–47.

Peirson, B. R. Erick. Wilhelm Johannsen's Genotype-Phenotype Distinction, Embryo Project Encyclopedia, ISSN: 1940-5030 (2012). (URL: http://embryo.asu.edu/handle/10776/4206). Accessed 10 March 2015.

Rheinberger, Hans-Jörg, Gaudillière, Jean-Paul (eds). Classical Genetic Research and its Legacy. The Mapping Cultures of Twentieth-Century Genetics, Routledge, London, 2004.

Sanger, F. et al. Nucleotide sequence of bacteriophage phi X174 DNA, Nature 265 (1977), 687–695.

Sismondo, Sergio. Maps and mapping practices. A deflationary approach, In: Gaudillière, Jean-Paul, Rheinberger, Hans-Jörg (eds). From Molecular Genetics to Genomics. The Mapping Cultures of Twentieth-Century Genetics, Routledge, London, 2004, 203–209.

The 1000 Genomes Project Consortium. A map of human genome variation from population scale sequencing, Nature 467(7319) (2010), 1061-1073. doi: 10.1038/nature09534.

The ENCODE Project Consortium. An Integrated Encyclopedia of DNA Elements in the Human Genome, Nature 489(7414) (2012), 57-74. doi: 10.1038/nature11247.

Verma, M. Personalized Medicine and Cancer, Journal of Personalized Medicine 2(1) (2012), 1-14. doi:10.3390/jpm2010001.

Waddington, C. H. The Strategy of the Genes. A Discussion of Some Aspects of Theoretical Biology, Allen and Unwin, London, 1957.

Watson, James D. Molecular Biology of the Gene. 3rd ed., Benjamin, Menlo Park, 1976.

Watson, J. D., Crick F. H. Molecular structure of nucleic acids, A structure for deoxyribose nucleic acid. Nature 171 (1953), 737–738.

Whiteford, N., Skelly, T., Curtis, C. et al. Swift: primary data analysis for the Illumina Solexa sequencing platform, Bioinformatics 25(17) (2009), 2194–2199. doi:10.1093/bioinformatics/btp383.

Witherspoon, D.J., Wooding, S., Rogers, A.R. et al. Genetic Similarities Within and Between Human Populations, Genetics 176(1) (2007), 351-359. doi: 10.1534/genetics.106.067355.

http://www.genome.gov/11006943. Accessed 6 March 2015.

http://www.genomicsengland.co.uk/the-100000-genomes-project/. Accessed 6 March 2015.

3

Who is the subject of genetic responsibility?

Angeliki Kerasidou, Cathy Herbrand, Malte Dreyer

The production, use and possible disclosure of genetic information about a person raises manifold questions of moral responsibility. We use the term 'genetic responsibility'[1] to indicate all the questions of moral responsibility that emerge in social contexts when there is a possibility of knowing personalized details about our genetic status. Genetic responsibility is thus about human differences. And it is about human self-understanding of these differences and the accountability that results from knowing or letting others know about them.

The reflections in the three parts of this exploratory chapter revolve around the question of who is the subject of genetic responsibility. Does genetic information carry a special normative force that imposes particular sets of responsibilities and requires specific action from the autonomous agent? And how should we understand autonomy in this context (part 1)? Is there something like a genetic human identity, which is bound to the whole genome? This question becomes urgent when mitochondrial genes are discussed. Does human genetic identity include the mitochondrial genome? And what would be the anthropological implications of including or excluding it (part 2)? In the final section of this chapter we turn to philosophical anthropology to helps us clarify *who* becomes genetically transparent.

---

[1] Genetic responsibility is not a new term: see for example: Weiner, 2011.

Responding to genetic information ethically[2]

The mapping of the first human genome in 2003 was hailed as the beginning of a new era in medicine. The hope was that genetic testing would make disease and health conditions detectable at a genetic level, opening the door to new, more effective methods of medical treatment and prevention. The newly discovered 'genetic transparency' did not confirm a strong genetic determinism; rather, it revealed the complex interactions between genetics, epigenetics and the environment, which together give rise to a person's phenotype. Geneticists discovered that only a relatively small number of diseases and conditions, such as Huntington's disease, Tay-Sachs disease, and cystic fibrosis, can be attributed to a single gene. Even in the case of genetic conditions with 100% penetrance, the expressivity of the condition can vary significantly between individuals.[3] Genetics did not expose a straight cause-effect relationship between genotypes and phenotypes, but it did uncover a huge range of information about the individual that until then had been hidden from view.

In the days before genetic testing was available, it was the clinical profile of a patient that would cause him or her to be defined as diseased or healthy. Laboratory tests, such as blood tests, X-rays and other scans were used to confirm the clinical diagnosis. Genetic testing has now been added to the medical armoury of preventive and diagnostic tools. Genetic testing can also be used to detect genetically linked health risks before an individual becomes ill. For example, people with a family history of a particular disease, such as breast cancer, could be offered genetic testing even before they develop any symptoms. Prior to the availability of genetic testing, doctors relied on the family history of the patient for this information. The genetic test can now confirm or alleviate fears that genetic traits have been inherited, and thus offer a more accurate picture of the individual's risk profile.

Non-communicable diseases, such as cardiovascular diseases, cancer and diabetes, account for 63% of all deaths worldwide. The numbers are expected to rise, mainly because more people are living

---

[2] This part was contributed by Angeliki Kerasidou.

[3] Miko, 2008, 137.

longer but increasingly unhealthy lives.[4] Environmental factors, such as pollution, and lifestyle habits, including smoking, eating, drinking and exercise, can have a significant effect on the increase or decrease in the risk of disease. People know that living in a polluted environment, smoking, eating foods high in sugar and saturated fat, and leading a sedentary life can have an adverse effect on their health. Public health educational policies and interventions in the past few decades have been quite successful in informing citizens about these risks.[5] Genetic testing can provide an extra level of information and make these generic public health guidelines more personal and perhaps more real by revealing how a person's own genetic make-up can affect their health prospects.[6]

As genetic testing has made health risk profiling, if not more accurate, then at least more specific and personal, how should people respond to this information? Do individuals have a moral obligation to respond by adapting or altering their lifestyle in order to influence the possible expression of a disease? Should people be held responsible for their bad health if they fail to adopt changes in response to their genetic risk profile, and if so, what form should this take?

The moral obligation to change

Doctors and public health professionals are hoping that genetics will open the door to better and more effective disease prevention strategies. The expectation is that if people are aware of their risk factors for disease, they will take responsibility for the lifestyle choices that are correlated with health and adopt changes that will improve their long-term health outcomes. Henceforth, '[p]reventing disease will also become the responsibility of the patient. He will know what [sic] the risks he takes if he smokes, over-eats or leads a

---

[4] Bloom et al., 2011.

[5] Naidoo et al., 2004.

[6] The company 23andMe had been advertising direct-to-consumer genetic testing to healthy individuals, urging them to find out how '[y]our genes – regardless of your health habits – may put you at higher risk for certain diseases associated with smoking and drinking. That can make the healthy choices you make that much more important.' Accessed on 18.08.2014.

sedentary life style. The risks will be personalized based on his own genetics.'[7]

There is a general intuition that it is reasonable to assign personal responsibility to people for their actions. As long as the action is freely and autonomously chosen, one should not be relieved of the cost of being a free and autonomous agent. Given that some of the current major killers, particularly in the western world, are lifestyle diseases such as obesity, Type 2 diabetes and cardiovascular disease, one could argue that a prudent patient will take the information regarding her risk of disease into account, and make the necessary changes in her lifestyle in order to help delay or even prevent the disease from occurring. As Daniel Wikler notes, '[i]n many cases, illness is not something that just happens to a person.'[8] The prudent person is thus the one who assumes genetic responsibility[9] before she or he is actually ill. People who take care of themselves, drink less, eat healthily, exercise regularly, sleep well, avoid stress and pollutants are likely to live longer and healthier lives. Acting with genetic responsibility will not only help her achieve a better and healthier life, but will also reduce the burden of disease to society as a whole.

It has been argued that genetics, by focusing on the individual, has played a significant role in promoting the concept of a person who is 'free yet responsible, enterprising, prudent, encouraging the conduct of life in a calculative manner by acts of choice with an eye to the future and to increasing self well-being and that of the family.'[10] People are expected to act in a particular, rational way and to exercise their autonomy responsibly. An autonomous agent is *ipso facto* responsible for her action, and subsequently for the outcomes of her actions. Autonomy is not only understood as a set of rights that the individual can exercise but also, as some have suggested,[11] a set of duties and obligations too.

It needs to be considered, however, whether we find it reasonable to hold people responsible for their health outcomes, and if so, how

---

[7] Steakley, 2012.
[8] Wikler, 2002, 47-55.
[9] Hallowell, 1999, 597-621.
[10] Rose, 2013, 341-352.
[11] Stirrat/Gill, 2005, 127-130.

society should respond to this. These two questions are closely related. The first pertains to the issues of how to interpret autonomy and responsibility. The second question relates to justice and fairness, mainly in the distribution of resources.

Autonomy and responsibility

Autonomy is one of the fundamental values in biomedical ethics, and has come to replace paternalism as the dominant paradigm in medicine. Instead of patients being expected to follow doctors' orders, the focus nowadays is on respecting individual patient autonomy. This has created the space for patients to determine their own notions of wellbeing, and has allowed them to reject doctors' recommendations that did not agree with their own idea of flourishing. The medical professional has moved from being a leader in managing a patient's life to the role of guide, the person who provides all the appropriate information and assists the patient in making an informed decision based on the patients' own understanding of good.

The principle of autonomy as a moral concept is usually traced back to Kantian ethics. For Kant, autonomy is the capacity to govern one's life in accordance with rational, and universalizable, principles. It is the manifestation of practical reason as expressed through the obligations a person has towards themselves and others. Yet this understanding of autonomy goes against the concept commonly used in bioethics: that of personal autonomy, where the individual is not acting in accordance with universalizable moral laws, but in line with her own personal reasons.

The rhetoric surrounding genetic testing, however, echoes a concept of autonomy that, as O'Neill observes[12], is much closer to the Kantian notion than that of personal autonomy. There seems to be the hope or expectation that, on discovering their at-risk status, people will make the required lifestyle changes and thus take responsibility for their health (for example, a person will start drinking less, exercising more, stop smoking etc.). This is perceived as the rational action, which should override personal preference. It

---

[12] Ibid.

seems, therefore, that genetic testing is not just a tool for deriving more health-related information about an individual. It also appears to carry a normative weight. The fact that risk of disease can be linked to the genetic profile of a particular individual with any degree of probability seems to have a morally action-guiding force. There is the expectation that the prudent and virtuous person would do 'the right thing', and try to prevent the disease or condition from occurring. Failure to act means that the individual should then also take the personal responsibility for her or his ill health.

Even if we accept that it is reasonable to require people to avoid actions that would put their health and wellbeing at risk, out of duty not just to themselves but also to others (e.g. their children) and to society as a whole, it is still questionable how universalizable this requirement can be. Consider this example: carrying the BRCA1 and BRCA2 mutations significantly increase a person's chances of developing breast and ovarian cancer. It has been found that bilateral elective mastectomy can reduce a person's chances of developing the disease.[13] Could we argue, therefore, that elective bilateral mastectomy is the appropriate and prudent response to this genetic risk? What if a double mastectomy would have a significant adverse effect on the mental and physical health of the at-risk patient and on her interpersonal and familial relationships? Does elective mastectomy in a situation where mental health is in danger or familial relationships are put under stress still carry the same normative force? Consider a second example. Somebody may carry a risk of developing respiratory conditions, and due to his socio-economic status he resides in an urban area with high air pollution. The person is advised to move to a less polluted area, but cannot afford it. If he moves out of the city, he will not only lose his job, but also the supportive network of friends and family around him, and his children will have to leave school. In his case, is moving to a rural area the right thing to do?

Although propositions such as '*all individuals should try to remain healthy and increase their wellbeing*' could work as a general rule, a more prescriptive proposition such as '*all individuals with a genetic risk of developing respiratory conditions should move to the countryside*' could not. This

---

[13] Metcalfe et al., 2014.

is because moving to the countryside might not be the right thing for everyone. We might describe a person who does everything possible to improve their clinical profile as a prudent patient; but prudence as a virtue is not about following rules; it is rather about exercising practical wisdom. Prudence, in Greek *sophrosyne*, refers to a person's ability to judge between virtuous and vicious actions in a particular space and time. Hence, when it comes to judging the appropriate response to genetic risk profiling, the prudent patient would be the one who employs reason and practical wisdom to make decisions that would increase health and wellbeing, rather than the person who will commit to following specific and prescriptive rules without further reflection.

Responsibility and Justice

The second issue to consider is accountability in its relation to justice. Should individuals be held accountable for their ill health, despite genetic testing having warned them about their risk factors? A prudent person might, after having considered all the available options, decide not to take any measures to prevent genetic risk from occurring. It seems logical to infer that this person should be held responsible for their actions, especially because their action was the outcome not of misinformation or lack of understanding, but rather of a conscious, possibly even careful and rational consideration of the situation. As mentioned above, autonomy, both in its Kantian and personal form, requires and demands the adoption of responsibility for one's actions. Otherwise the action cannot be considered autonomous.

Responsibility, however, also necessitates causation. We need to be able to demonstrate that a person's action has led to a particular outcome for which they should take responsibility. Many of the conditions for which genetic risk can be predicted, such as cancer, diabetes or respiratory and cardiovascular diseases, are multifactorial. Whether the individual who carries the risk will go on to develop the disease depends, as mentioned above, on genetics and on complex environmental factors, such as nutrition, pollution, and the overall physical and psychological condition of the individual. Hence, although a correlation can be shown between a person's lifestyle choices and their health outcomes, a strong causation cannot always be established.

Yet even if we cannot ascribe causal responsibility, perhaps we can ascribe *moral* responsibility to individuals who decide not to change their habits and life in order to prevent genetic risk. A person has a moral responsibility towards himself and those around him to look after his health and wellbeing. Responding to information about genetic risk by changing one's lifestyle is one way of fulfilling this responsibility. Let us look at our earlier example of the man with the risk of developing respiratory disease. He knows that he is at risk for respiratory disease, but he also knows that if he moves to the countryside, he will lose his job and his children will have to leave school. Of course, remaining healthy is an important factor to his future and wellbeing, and that of his children, but other things are important as well, such as a steady income and education. He may decide, after considering his situation and obligations, that it is best to stay where he is and risk increasing his chances of suffering from a respiratory disease later in life. His decision, then, is autonomous and informed by reason. Regardless of whether one agrees or disagrees with his decision, one cannot accuse him of not understanding his moral responsibilities.

We can counter-argue, however, that even if the man in our case does not deserve to be blamed for his decision to stay, he definitely does not deserve to be 'rewarded' either, especially if his decision is proven to be wrong and he does develop a serious respiratory disease. Shouldn't he be held responsible for his health outcome? And shouldn't the healthcare providers take this into account when making decisions regarding health care provision? It would be unfair to treat other people, who decided to take the advice of the health professionals and relocate to healthier areas, the same as those who knowingly take the risk and ignore the doctor's advice. One way of addressing this unfair situation is to give people who ignore medical advice lower priority in the allocation of resources, or even to make them pay for their own healthcare.

The discussion about ascribing personal responsibility for health is a complex but not a new one in bioethics and public health ethics.[14] Today's debate on genetic responsibility could be seen as an extension of the earlier one on personal responsibility and distributive

---

[14] See: Wikler, 1985; Firser, 1982; Crawford, 1977; Veatch, 1981; Harris, 1995.

justice in healthcare. Given the inability to establish clear causal and moral genetic responsibility for ill health, using lifestyle choices as a way of limiting people's access to healthcare resources would create more injustice rather than it could eliminate.[15] People who suffer ill health because of their lifestyle are already effectively punished for their wrong decision. As Harris notes, any attempt to limit their access to healthcare would only be seen as an additional form of punishment, and this would be unjust.[16] Even in cases where a strong causal connection could be established between genetic risk and health outcome, it still remains questionable how society ought to respond. It is indisputable that society allows, even supports, some personal risk-taking behaviour (e.g. competitive sports). However, we lack a criterion that would allow us to decide how much risk individuals are allowed 'for free', before they start becoming a burden to others and to society.[17] When it comes to health, the acceptable risk threshold should be much higher based on the premise that health is a primary good,[18] and necessary for individual flourishing. A just society committed to protecting and promoting fair equality of opportunity and to supporting liberal values of self-determination and the right to pursuit a meaningful life ought not to limit the access people have to this good.[19]

Conclusion

We have argued that the ability to detect genetic risk has created a new category of individual, the 'at-risk individual', a person who is not yet ill but has potential to develop a disease. The potential to develop a disease is not something new – taking a family history was a way of assessing health risk before genetic testing was available. However, genetic testing has made the assessment of risks more 'objective', and also more personal, more individual. Hence it has

[15] Harris, 1995.

[16] Ibid.

[17] Wikler, 1987, 11-25.

[18] Daniels, 1985.

[19] Minkler observes that the argument of personal responsibility is commonly used by governments in order to justify cuts in necessary health and social services programmes. See: Minkler, 1999, 121-141.

normative force ascribed to it. Although any genetic risk still remains probabilistic, the prudent patient is expected to take steps to mitigate it, for example by making lifestyle changes.

Nevertheless, this paper challenges the view that genetic risk profiling carries a special normative force that would necessarily require lifestyle changes. The principle of autonomy and the virtue of prudence do not necessitate that a person follow specific and prescriptive rules; and so the prudent and autonomous patient is not necessarily and in all situations the one who decides to undertake lifestyle changes. Prudent and autonomous behaviour demands that a person takes all the information into account, and considers the specifics of their situation before making a decision. Sometimes the decision would be to make lifestyle changes, for example stop smoking or avoid air pollutants by moving to a rural area, but at other times the prudent thing to do will be to carry on as before and take the risk. The prudent and autonomous patient remains the one who can exercise rational thinking and practical wisdom.

Finally, we argue that holding people responsible for their health and using this attribution of responsibility as a way of making resource allocation decisions would increase injustice and unfairness in society. As mentioned above, health is a significant factor in ensuring equality and fair opportunity for all. Rationing people's access to healthcare based on lifestyle choices would work as a double punishment by limiting or even denying to those already worse off (because of their poor health) their chance of recovery.

Now we turn to a different aspect of genetic responsibility. Would it be responsible to *change* one's genes, or the genome of any individual, in order to prevent a disease? Is it even conceivable, given that the identity of a person might be changed by changing the genome? Germ-line modification has been considered illicit for these reasons. Therapy should treat but not change the person. However, is a complete ban morally defensible? This question came up again in the context of *mitochondrial* DNA, which is contained in each of the human cell, outside the nucleus.

Nuclear transfer and mitochondrial DNA: Changes in human self-understanding[20]

Besides creating the new category of 'at-risk individuals', progress made towards 'genetic transparency' has also generated new biological distinctions and categories, namely those of 'nuclear DNA' and 'mitochondrial DNA'. This distinction has become especially salient in the recent debates regarding mitochondrial replacement (MR) techniques aimed at preventing the transmission of maternally inherited mitochondrial disorders. Not only have these debates challenged the definition of germ-line therapy and the ban on germ-line therapy, they ultimately also question current conceptions of 'genetic identity'. In what way is our idea of *who we are* connected to the individuality of our genome?

The discussion was raised by considering whether mitochondrial diseases, which can cause life-threatening syndromes, could be prevented by new cell reconstruction techniques involving mitochondrial replacement. Mitochondrial disorders are caused by mutations in the cell's mitochondrial DNA and can create very severe disorders and various body dysfunctions, for which no treatment is yet available. In addition, women carrying the mutation are likely to transmit the disorder to their biological offspring. New mitochondrial replacement techniques would enable some of these women to have a healthy biological child, by transferring the nucleus of the affected mother's egg (containing the nuclear genome) into a healthy enucleated donor egg cell. This technique has been called 'maternal spindle transfer'. A similar intervention can also be applied after fertilisation in an early embryo: this is called 'pronuclear transfer'.

The striking novelty of both maternal spindle transfer and pronuclear transfer is that the conceived child will inherit DNA from three individuals: the nuclear genome (which is thought to determine all our individual physical characteristics) from both intended parents (99.9%) and also, albeit to a much lesser extent (0.1%), the mitochondrial DNA from the oocyte donor. Moreover, this donated

---

[20] This part was contributed by Cathy Herbrand. A previous version has appeared in a special issue of Law and the Human Genome Review. See: Herbrand, 2014.

mitochondrial DNA will be transmitted over the generations through the maternal line. Pronuclear transfer techniques have been first developed in the UK by a research team in Newcastle upon Tyne, who successfully applied these techniques to human embryos.[21] After public and parliamentary debates, these techniques are now legal in the UK. Researchers hope that the first healthy baby conceived using these techniques will be born by 2016.

These techniques nevertheless still raise several fundamental ethical questions, in particular whether or not MR techniques that combine a given set of nuclear chromosomes with different mitochondria (and their mitochondrial genomes) from a donated egg cell should be regarded as germ-line gene therapy. Germ-line modifications generally refer to genetic modifications of the gametes or the early embryo, which will be passed on to subsequent generations, unlike 'somatic' gene therapies that affect only the patient concerned.

A key issue for MR techniques is therefore whether or not to conceptualise them as germ-line modifications. MR techniques do indeed modify an individual's mitochondrial genome by replacing the mother's mutated mitochondrial DNA with that of the egg donor. The donor's mitochondria DNA will then be passed on to subsequent generations along the maternal line.

Current conceptions of that matter are strongly divergent and have varied ethical and legal implications. As we shall see, one of the main elements of divergence is which criteria are relevant to ascertaining the conceptual status of these techniques. Another crucial decision is determining the ethical acceptability of these techniques following such conceptions in order to allow or to prohibit them.

Ethical positions discussed so far

Overall, the main arguments and stances that have appeared so far can be theoretically mapped into four major ethical positions:

a. MR techniques are not germ-line therapy and are ethically acceptable

---

[21] Craven et al., 2010, 82-85.

Some experts, including several of the UK researchers in the Newcastle research team who have developed MR techniques, do not regard them as germ-line therapies.[22] They argue that germ-line modifications initially referred to changes in the nuclear DNA, which is thought to determine all our individual characteristics. In that respect, MR techniques leave the nuclear DNA intact. According to these researchers, while there are different kinds of risks in modifying the nuclear genome because of its complexity, mitochondrial DNA is much simpler and exchanging it in the case of mitochondrial disorders can be compared to changing a defunct battery in an electrical clock. Moreover, there is a significant distinction between nuclear DNA and mitochondrial DNA in terms of biological function and composition.

For these various protagonists, the fact that MR techniques cannot be considered germ-line therapy means that these techniques should not be treated as such and prohibited. This stance favours permitting MR techniques for treatment, while maintaining the ban on germ-line therapy strictly defined as modifications of the nuclear DNA.

b. MR techniques are not germ-line therapy but are ethically unacceptable

Interestingly, some protagonists acknowledge that modifications of the mitochondrial genome is indeed different and not as problematic as modification of the nuclear genome, but they argue that allowing the former will inevitably lead to allowing the latter at some point in the future. It could also open the door to the acceptance of human enhancement. They therefore put forward a slippery slope argument to prohibit MR techniques.

This ethical position is also followed by opponents to the techniques who cite other kinds of reasons for prohibiting them, such as the impossibility of obtaining the informed consent of the resulting child, the availability of reproductive alternatives, or the significant amount of financial and human resources needed to develop and implement these techniques. However, these arguments are not directly related to germ-line modifications as such.

---

[22] See: North East England Stem Cell Institute, 2008.

c. MR techniques are germ-line therapy and are ethically unacceptable

Conversely, many people do regard MR techniques as germ-line therapies. A first reason given defending this position is the fact that they involve genetic changes that will be transmitted over the generations. This means that not only the person conceived using these techniques will be affected by them; the future generations will also incur some risks in terms of health, and they will not have the same mitochondrial inheritance as their maternal ancestors. As it is difficult to assess the possible short- and long-term effects of these techniques without observing them over several generations, some argue that it is not reasonable to allow such techniques. This stance appears to relate to the precautionary principle.

Apart from their intergenerational effects, some people also consider MR techniques to be germ-lime therapies because they would alter genetic inheritance as a whole. The genetic profile of the resulting offspring will in fact not be the same if their maternally inherited mitochondrial DNA has been replaced by that of an egg as if it had not. This argument is then used to present MR techniques as ethically unacceptable on the grounds of the same principles previously used to prohibit germ-line therapies,[23] such as the individual's right to an open future, the need to preserve the individual's unique characteristics and genetic inheritance, the risk of using them in the future for human enhancements, the preservation of the 'order of nature itself', and so on.

d. MR techniques are germ-line therapy but are ethically acceptable

This relatively novel and unusual ethical position was notably taken by the Nuffield Council on Bioethics (NCB). In their recent on MR techniques, the NCB Working Group indeed argued that MR techniques represent a form of germ-line therapy that modifies the mitochondrial genome and will be transmitted over the generations.[24] In so doing, they follow the criteria previously put forward by A. Bredenoord, a bioethicist who disagrees with making an ethically significant distinction between modification of the nuclear DNA and modification of the mitochondria DNA. Among other reasons, this

---

[23] Munson et al., 1992, 137-158.

[24] Nuffield Council on Bioethics, 2012.

distinction is problematic as some uncertainties remain regarding the biological functions of mitochondria and the nucleo-mitochondrial interactions. But more significantly, she argues that both affect the 'qualitative identity' of the future person. Indeed, some aspects of the person conceived using MR techniques will definitely be different from what they would have been, had the techniques not been used.[25] Not only will the person's mitochondrial genome be altered, even though in a minor way, but the person's self-conception will also differ significantly from what his or her self-conception would have been as a person developing mitochondrial disorders. In addition, this person's self-conception might be affected by knowing that s/he has been medically conceived including the donor's genetic contribution.

According to the NCB Working Group, MR techniques can also been regarded as altering the 'numerical identity' of the resulting offspring because 'the inclusion of a donor's mitochondrial genes and minimisation of the proportion of maternal mitochondrial genes could make such a very significant difference to the resulting person's life that they could be said to make them 'a different person'.[26] While the NCB Working Group therefore definitely considers MR techniques to be germ-line therapies, they nonetheless add that, 'some changes to the mitochondrial genes have germline effects that are different from the germline effects of changes to nuclear genes. Differences include that [MR techniques] are not intended or known to affect nuclear genes; they aim to make no changes to the donor's mitochondria; and only women born from these techniques would be able to pass the changes on to their children.'[27] This means that although germ-line modifications should primarily be defined in terms of their impact on the future person's identity, the Working Group believes that different types of germ-line modifications could be acknowledged according to the genome targeted, the nature of the genetic manipulation (replacement or specific modification), and their scope (impact on all resulting offspring or only some of them). This stance leads the Working Group to imply that different types of legal

---

[25] Bredenoord et al., 2011, 97-100.

[26] Nuffield Council on Bioethics, 2012, 55.

[27] Ibid, 58-59.

regulation could also be required, even though this question needs further debate.[28]

Acknowledging that MR techniques are qualitatively, and possibly numerically, identity altering does not mean, according to Bredenoord and the NCB Working Group, that MR techniques are ethically unacceptable and should be prohibited. Indeed, such effects, according to them, are not specific to MR techniques and, more importantly, they do not harm the resulting child's 'right to an open future', the idea that the child's future options should be kept as open as possible until s/he is capable of making her/his own decisions. It is rather the contrary, as these techniques will enable the child to avoid very serious disorders that would compromise her/his life and projects.[29]

Genomic and personal identity

Mapping the positions and arguments about mitochondrial transfer techniques as germ-line modifications enables us to indicate the four major options that may guide the regulation of MR techniques here. It also shows the complexity of this issue by pointing out the difficulty of determining the relevant criteria for categorising these techniques into germ-line alterations or not. Should the biological distinction between nuclear DNA and mitochondrial DNA be taken into account, or should it rather be the modification of the (qualitative or numerical) identity itself? While it seems risky to rely on a biological distinction, which can vary according to the current state of scientific knowledge of genetic functions and functioning, it also appears challenging to determine theoretically the ways in which MR techniques might alter the essential characteristics that many believe influence our identities.[30] Deciding whether or not such a distinction can be sustained despite such complexities clearly requires careful consideration.

The issue then is whether such distinctions justify differentiated regulation. Can we challenge the current ban on germ-line therapies? These techniques indeed bring into question the broader issue of

---

[28] Ibid.

[29] Bredenoord et al., 2011, 97-100.

[30] Bredenoord et al., 2008, 669-678.

what defines germ-line modifications. There is obviously a need for more precision in this respect, a clarification of the reasons *why* germ-line modifications are considered illicit, and then possibly the creation of different kinds of sub-categories. This might also be the opportunity to further clarify what human identity means in terms of the genome, and more specifically the mitochondrial genome; what the different parts of the genome contribute to 'human' and 'individual identity'. Then we can discuss on a more secure base how to consider 'identity changes' ethically. How do these distinctions challenge the definition of personhood by localising our 'important features' within the nuclear DNA? Should we look at our genetic inheritance as a whole? Should our identity be reduced to this DNA as it determines certain phenotypic characteristics, excluding the mitochondrial DNA, mechanistically understood as just a 'battery' producing the energy of our cells? On the other side, increasing effort has been put into the search for ancestry through the analysis of just these maternal mitochondrial genes. Indeed, mitochondrial inheritance has recently been given growing importance for tracing maternal line ancestry and finding out about one's personal and cultural 'identity' (Nash, 2004).

As we have seen, the interpretation of the ethical significance of the genome (and of alterations of the genome) touches fundamental issues of how we understand human identity. These questions are discussed in philosophical anthropology. The next section deals with them.

Homo geneticus? Considerations from philosophical anthropology[31]

We often understand anthropology to mean a particular scientific discipline that investigates ancient human bodies. However, we would like to use the term 'anthropology' in the text that follows in a much broader sense, to describe the philosophical enquiry into the concepts of 'human', and 'human being' (in German: *Mensch*). Why are such considerations relevant to genetic transparency? All species

---

[31] This part was contributed by Malte Dreyer.

have a genome, so the possession of a genome is clearly no human speciality. But the 'human' genome might define humanity in some way. As Nelkin and Lindee[32] have demonstrated, the cultural images and symbols of the genome have a very strong metaphysical undertone, suggesting the genome to be the essence, even the soul, of a human being. Within such a naturalistic view, the anthropological significance of the human genetic make-up seems obvious. But is this defensible?

Many parts of the genetic make-up of human beings undoubtedly belong to the properties that the human shares with other species. But it should not be overlooked that in handling genetic data we differentiate between human data and those of animals or plants. Even if we only rarely justify this difference or explain it conceptually, it is a significant part of our common practice and policies. Some of the differences are easy to explain. For example, when determining, processing and publishing genetic information about animals and plants, different legal rules and ethical standards apply than when handling the genetic information of humans. We consider other humans as individuals with human rights, including a right to privacy. Such rights are not granted to plants and animals. In these issues the protection of genetic information as 'personal information' is based on the legal recognition of others as persons with rights. But could the question also be turned around? Is the basis of legal recognition of these others their 'human' genome? In order to understand the meaning of such questions we can draw on the philosophical anthropologies of Ernst Cassirer and Helmut Plessner.

The anthropological question

Explanations of what makes a human 'human' are given by anthropology. Anthropology is not really an independent scientific discipline. It is much more a description of the questions shared by a multitude of disciplines. In the early 1970s a multi-volume, interdisciplinary series on anthropology was published in Germany. Its declared aim was to provide an overview of current

[32] Nelkin/Lindee, 1995. See also chapters 1 + 2.

anthropological research.[33] A quick review of the literature shows that the contents of the published volumes are in many ways representative of anthropology in 'Western' thought. It shows that engagement with that which is specifically human leads in very different directions, in terms of both method and content. In addition to volumes on biological anthropology, there are others on social anthropology, cultural anthropology, psychological anthropology and philosophical anthropology. Although these different disciplines, each with its own method, all appear to be broadly directed towards questions about the specifically human, a closer look shows that in fact the object of their interest differs markedly.

Thus, biological anthropology principally investigates physical human characteristics in comparison with animal and plant organisms, while social anthropological studies seek the specifically human in the relationships between human individuals. In the case of biological anthropology, the *differencia specifica*[34] are located in the body of the individual, while in social anthropological terms they are defined as something intersubjective and not materially determined. It is certainly beyond dispute that both physical and non-physical properties characterise the human being, but there is disagreement about which of these two groups of properties is primary. It is however essential to the topic under discussion here. If we speak of the human being as a genetically constituted entity, thus understanding the human *essentially* as a collection of particular physical properties, the practical, political and moral questions (discussed in the next chapters) may be posed quite differently from how they would be if we view particular physical properties merely as the *necessary conditions* for humans to be differentiated from animals and plants by cultural practices that cannot be reduced to their biology, such as the formation of political communities or the use of symbols. We cannot at this point list all the reasons for or against searching within the natural characteristics of the human for the *differencia specifica*. But what is important about the underlying concept

---

[33] Gadamer et al., 1972.

[34] 'Differencia specifica' means here the feature that distinguishes the human being from non-human entities.

of the human is that it is characterised by tension between the biological and the practical (ethical, legal, political, social) dispositions. Correspondingly, 'genetic transparency' should both describe a perspective on the human body and also indicate a practical dimension that is not manifested in the individual body. Although it is a property of the material body of the human, it also reflects relationships to the self and to other individuals.

Anthropological scepticism

For a long time philosophical anthropology assumed that many of the properties corresponding to the human, i.e. the 'anthropogenic characteristics', are unchanging over time and across cultures. For example, it has been assumed that human beings have distinctive abilities to use symbols,[35] to develop consciousness, to think rationally,[36] to use tools,[37] to recognise themselves, to develop economies, to perform particular practices – such as play[38] – or to act in other ways that form social, economic and/or political communities.[39] Assumptions of these kinds were however subjected to fundamental criticism as early as the second half of the 20th century.[40] For one thing, the historical and cultural fluidity of that which is considered human, as seen throughout history and across different cultures, poses the fundamental question of whether there are any invariant human properties at all. Indeed, a glance at the history of philosophical anthropology reveals that many statements about properties that had been assumed to characterise humans independently of all historical and cultural developments were rooted in precisely these historical and cultural contexts. For example, Arnold Gehlen's much-discussed early 20th-century definition of the human, as a being that compensates for its physical deficiencies through cognitive competences by emancipating itself from its

---

[35] See Morris and his remarks on 'Humanistic Implications of Semiotic". Morris, 1966, 57-59.

[36] Kant, B312-B316.

[37] Gehlen, 1940.

[38] Huizinga, 1938.

[39] Aristotle, Pol. 1253a1-15. See also: Arendt, 1958, 181-187.

[40] Foucault, 2008, 351-354.

natural environment using complex technologies, is a statement that can be ascribed to a certain technological euphoria of the time.[41] Very likely, today's recognised anthropogenic features will one day also be shown to be just as dependent on dominating theoretical fashions and ideologies. Furthermore, the results of primate research require continuous corrections to the list of anthropogenic features. For example, all the above-mentioned entries to the list of distinctively anthropogenic features have been effaced by observations of animals that use symbols, recognise themselves in the mirror, trade, show signs of rational action, use tools, play, or form societies.[42] Distinguishing between animal and human by using a list of these and similar properties could only be done at the price of the additional assumption that we have to differentiate between a *specifically animal* and a *specifically human* way of using tools, or a *specifically animal* and a *specifically human* way of forming states etc. The question of what the distinctive features are that determine a human is thus not answered, but only displaced into another area. Instead of defined properties it now aims for a mode that is shared by all entries on the list, or for a particular way in which properties are displayed.

A further fundamental problem of anthropology is the use of the generic singular, 'the' human. The use of this term would be ethically problematic if it encompassed only those exemplars of a species that possess the anthropogenic features in reality. This would exclude individual beings that may not currently possess these properties but have possessed them or may acquire them in the future. Infants, patients in comas, or deceased persons would thus be excluded from the class of humans. If we wish to avoid this counterintuitive and discriminatory result – as Foucault concludes – the phrase 'the human' may not be used either descriptively or prescriptively as correct word usage.[43]

In view of this and similar problems it appeared tempting to many philosophical anthropologists to withdraw to the assertion that 'the

---

[41] For Adorno's critic on Gehlen see the famous discussion between Adorno and Gehlen: https://www.youtube.com/watch?v=a9GB_XGnKyw.

[42] Sommer/Hof, 2010; de Waal, 2009.

[43] Foucault, 2008, 351-354.

human' was unfathomable.[44] But in times in which the specifically human is subjugated by massive reinterpretations caused by the challenges of modern biotechnological developments, if we hide behind a thesis of unknowability we give the field over to those who define the human unambiguously as an exclusively biological being.[45] At this point we should not abandon the thought that as well as a material, physical dimension, 'human' possesses a further dimension of incorporeal significance. However, before we go into the relationship of these dimensions of the human, we must first analyse the *logical form* of the term 'human' before a premature narrowing to particular features of *content.* In the next section we suggest how to understand the logical form of this term by presenting the theories of Plessner and Cassirer.

The 'animal symbolicum': from human being to human image

So far we have only discussed empirical properties. In this case, however, listing empirical properties that could determine the specifically human does not help provide an adequate definition of the term 'human'. We need to take into consideration the *form* of the term as well. An outstanding attempt at such a definition of form is made in Ernst Cassirer's anthropological cultural philosophy. What is special about this is that it reconstructs the means with which philosophical anthropologists work, i.e. their language as a medium of specifically human expression. The human being, in Cassirer's eyes, is an animal that uses symbols, that expresses itself exclusively through symbols, and at the same time gazes through a veil of symbolic mediations at his or her own reality and on herself. In addition to religion and myth, these symbol systems include philosophical language and science: the knowledge systems with which we refer to the human as a genetically transparent being.

Cassirer's "Philosophy of Symbolic Forms"[46] and his later "Essay on Man"[47] aim to provide a comprehensive, cultural philosophical foundation of anthropology. In contrast to the anthropology of, for

---

[44] Arendt, 1958, 181-187.
[45] Dreyer/Rehmann-Sutter/Erdmann, 2014, 260-267.
[46] Cassirer, 2010.
[47] Cassirer, 1948.

example, Arnold Gehlen or Huizinger, which were developed at the same time, Cassirer does not attempt to justify the specifically human by means of capabilities *the content of which is precisely defined*, but to define the general *form* of human apprehension of the world and human expression. In doing so, he forms the *differencia specifica* with recourse to what was in his time a prominent biological theory, indicating that anthropological theories always have one leg in the field of natural sciences. The theory to which Cassirer refers originates from Johann von Uexküll. According to Uexküll, every organism relates to its life-world (*Umwelt*) in a way that is specific to its species. Uexküll makes a fundamental differentiation between two orientations of relationship to the world. He uses the term "receptor system" (*Merknetz*) to describe classes of relationship to the world in which an organism is harnessed through sensory, haptic, visual and other stimuli. In an "effector system" (*Wirknetz*), the same organism is woven into its life-world through the possibility of reaction.[48] The two kinds of integration are interrelated, in a way that can be reconstructed from the receptive side: the effect an organism has on its environment depends on how the world appears to it. That a bird of prey is able to hunt rodents depends essentially on its ability to see them from a great height. Due to their sensitive noses, dogs perceive their environment through the smells that surround them to a much greater extent than other organisms, and thus they have the ability to follow scent trails or to identify other members of their own species without visual contact. This dependency on species-specific ways of reacting to the world becomes particularly clear by comparing the remotest coordinates on the enormous range of organisms living on our planet – for example, an insect with an elephant. In such a comparison it is not difficult to see that visual and haptic impressions, as well as ideas of duration or spatial extension, differ between species. This observation leads Uexküll to the conclusion that it is not just the organisms and their ways of perception that differ, but they also differ according to *the worlds* in which they live. The world of flies, Uexküll would state in a constructivist exaggeration, is *a different world* from the world of elephants.

[48] Cassirer, 1926, especially chapter III.

Anthropologically we are thus not asking about the human in the world, but about the world of the human.

In comparison to the previously mentioned animal organisms, the human – according to Cassirer following Uexküll – has an additional reference system, which compared to the pure receptor and effector systems *Wirknetz* and *Merknetz* opens up a qualitatively new dimension of access to the world. Humans, according to Cassirer, communicate their specifically human reality through the use of *symbols*.[49] The term "symbol", in this context, means not an iconographical sign in the narrowest sense of the word, but much more fundamentally the medium of each form of human expression in which a meaning is created through the connection of a signifier with the object it describes. Since Cassirer's definition initially only cites a *type of classification* and refrains from naming concrete symbolic contents, he also speaks of *symbolic forms* rather than just symbols: "Under a 'symbolic form' each energy of the spirit should be understood through which a spiritual meaning or content is joined to a concrete sensory sign and is inwardly adapted to it."[50] Almost all human practices and their results should therefore be regarded as "symbolic" if they make reference to another semantic content than that given by their mere presence. This reference is not a causal but a logical one. Cassirer illustrates this logical relationship through a comparison of the symbol with the signal: while the latter makes reference to its cause in the way that a locomotive whistle does, the classification of a sound and the object it designates in human language is at one and the same time characterised by convention, malleable, and subject to dynamic changes.[51] This sometimes very productive play with meanings takes place not only in the field of human language, but is also encountered in art, in technical artefacts, in religion and many other aspects of culture.[52]

---

[49] Cassirer, 1948, chapters II + III.

[50] Cassirer, 2003, 79.

[51] Cassirer, 1948, 36-37. Emotional reactions can cause expressions which are similar to signals. In this spirit Cassirer differentiates between emotional and propositional language. See: Ibid, 30. See also Chapter VIII.1.

[52] Cassirer differentiates several forms of culture in which a definition of 'humanity' takes place: myth and religion, language, art, history and science.

In addition to this productive dimension of the symbolic there also exists a receptive dimension. Humans perceive their world through symbolic communication as much as they design it. This symbolic reshaping of human reality can also extend to objects in nature – for example in the way a mountain is seen as the home of the gods, or the human body as an information carrier. The most fundamental categories of our perception of reality are woven into the "symbolic net" (*Symbolnetz*). Even the ways in which space and time become accessible to us are determined, in Cassirer's cultural philosophical perspective, by the reception and production of particular meanings of space and time.[53] By contrast, animals have only a view of the present of space and time, which is communicated exclusively through their sensory organs. In the course of our cultural development, complex constructs such as Euclidean or relative space have entered our perception. The influence of our semiotic activity on our self-awareness becomes particularly clear in the example of time. In addition to the idea of duration, which is formed by waiting for prey or the change from day to night, humanity is additionally characterised by the idea of *history*. As the example of family (hi)stories shows, such stories – insofar as they also concern kinship and heritable properties – have a biological dimension as well.

Cassirer's anthropological considerations are noteworthy in that they do not deviate from the framework of symbolic communication. Thus the *differencia specifica* illustrated by the example of space and time can only be explained through the medium of symbolic form. Even the assumption of a non-human perception of time that has not yet been troubled can only be articulated within a specific symbol system.

In terms of human culture, Cassirer differentiates between various symbol *systems*, including religion, myth, language and science. These characterise human reality in an equally fundamental but qualitatively quite different way from the biological determinants of sensory perception. Comparing the biological composition of the human being and his or her ability to use symbols, both differences and commonalities become evident. They share the effect of forming the world: just as an organism that can see has a representation of a

---

[53] Cassirer, 1948, 42-55.

different world from one that cannot see, a world accessed against a background of mythical, religious, aesthetic and scientific interpretations has a fundamentally different form from a world that cannot be inserted into these symbol systems. Nevertheless, symbol systems differ from the biological determinants of human perception in one essential point: symbol systems are the results of socio-cultural practices and can thus be designed and developed. This requires that we be able to describe and criticise them. As is to be expected, this reference to our symbol systems is only linguistic, scientific, mythical; in short, it is possible only through the use of symbols. The human being is as unable to escape from the "net" of symbolic references as animal organisms are unable to liberate themselves from the sensorily structured reference systems of their perception of the world. If we practise anthropology and seek what is specifically human, we likewise find ourselves always already in a net of meanings and interpretations.

If we now relate this quite general finding to our topic and attempt to apply Cassirer's considerations to the genetic perception of the human body, we can state four things. (i) Our embodiment is part of our human reality and is thus symbolically communicated. The body is already equipped with a particular meaning, which is handed down culturally and can become the object of critique. (ii) In terms of cultural anthropology, we can then ask in what form and with what results this takes place. This makes apparent that the perspective on our biological endowment is dependent on the prevailing scientific opinion. Today, the human body appears to us as the carrier of genetic information and risks, is described as a manifestation of properties that endure over generations, and is itself a link in a chain of transmission of codes. (iii) The human body is not just the object of scientific symbol systems, but is open to representation using the vocabulary of other cultural fields, including religious, mythical and aesthetic connotations. Describing the interdependencies of these different representations is the task of a detailed cultural anthropological investigation.[54] (iv) What is essential for the anthropological implications of genetic transparency is the observation that the object we access genetically has never presented

---

[54] Wulf, 2011.

itself to us in its natural state, even before the invention of the high-throughput sequencer, but was always entangled in a net of interpretations in the laboratory, the clinic, the church, art and so on. Furthermore, based on Cassirer's considerations we may not assume that these biotechnological developments allow us to apprehend the *true* and *real* nature of the human body. Much more, the nature of the human body becomes understandable as changes in the symbol system of science, an understanding which in a novel way colours the veil through which we contemplate ourselves, our bodies and the world around us.

What does this kind of constructivist reconstruction contribute? A symbolic philosophical anthropology has several advantages. For one, it does not prioritise any of the many possible perspectives on the human body, and thus prevents the sort of practical demands that would arise if the handling of genetic information is equipped with the virtual power of 'natural' conditions. What demands should be made of genetic counselling, for example, or how patients' rights should be reformed, cannot be decided by reference to the objectified body alone. If we discuss how to react practically (legally and socially) to the availability of personal genetic information, we cope with symbolically communicated representations of the human. There is no *one* single correct reaction to genetic screening based on a generally valid definition of the human, because the *differencia specifica* of the human is an *animal symbolicum* that always obstructs a free view of human nature.

Second, this approach allows the possibility of open communication about different anthropological settlements. Symbols are always components of an open communicative process. As a meaningful symbol, the human is the object of a shared understanding. A position inspired by a cultural anthropology such as Cassirer's opens up the freedom to describe different human images, to distinguish grounds for or against them, and to make recommendations about changing anthropological conditions.

A further point is significant. As discussed previously, symbolic systems do not have an exclusively linguistic nature. Technical artefacts, rituals and other products of practice can be understood within the framework of Cassirer's theoretical setting as "symbolic". In this respect even technical artefacts or works of art have a symbolic dimension. Thus, non-linguistic forms of human expression can also be interrogated for their anthropological content. Visual arts from a particular period can be investigated to identify which concept of the human is sedimented therein. For example, in comparing Renaissance painting with the iconography of the Middle Ages, it can be shown that at the transition between these two periods the picture of the human is recontoured through the conflict between religious conviction and scientific discoveries.[55] By analogy, the shift in diagnostic apparatus from the stethoscope to genetic testing can also be accorded a symbolic dimension and an anthropological relevance. We can analyse objects, practices and the knowledge they produce as meaningful symbols in human self-expression. Precise analyses of their anthropological content need further investigations. By now we suppose that genetic testing changes our view of human bodies in at least one important aspect: whereas conventional diagnostic methods are related to individual bodies and their physical condition, genetic testing evokes the imagination of transferable and shared information. The terminology we choose to explain and understand the operating principles of genetics is similar to the language we use to describe digital data transfer: sequencers 'read data', 'produce datasets', persons 'carry genetic information' and inheritance is described as a sort of 'copying'. The reproduction and distribution of data raises questions of ownership and copyright. As a result we discuss whether the owner of the data medium, the person who produced the data or those whom the data concerns are legal owners of genetic information. As far as sequencers are data readers, *homo geneticus* is a data carrier.

Self-objectification

With the ability to use symbols, the human enters into a meaningful, mediated relationship to the world in which he or she develops his or her reality. Helmuth Plessner, a contemporary of Cassirer's, made the reflexive ability of humans to produce references the centrepiece of his anthropological theory. With a spatial metaphor he describes the standpoint that we ourselves take up and that is claimed to be characteristic of the human as "eccentrically positioned".[56] In

---

[55] Petrarch's ascent of Mount Ventoux is usually viewed in this context. See Ritter, 1974, 105-141.

[56] Plessner, 1980, 360-382.

contrast to the merely "centrical" animal,[57] able to locate itself in its environment and relate to itself, the human also has the ability to behave in a formative way towards self-reference. Only the human, Plessner believed, has both the task and the capability to make his or her self-relation into the object of a reflexive reference. With these considerations, he has provided a concept that supplements Cassirer's cultural philosophical approach with natural philosophical components: humans are beings that are capable of self-relation without having first to be emancipated through cultural and technical means from their natural context. Already *as humans* they take up a distance from themselves. This structure of human self-knowledge is a condition, and not the result, of the numerous self-descriptions that we know from art, politics, science and everyday life. Yet as these examples show, the eccentric position, although it is given to humans by nature, is always a construction, a reflected standpoint.[58] The human is thus, by nature, a cultural being, or 'naturally artificial'.

Something else becomes clear if we look at the possible diversity of human self-relations: the eccentric position does not describe a single, precisely defined standpoint, from which the human or the self can be exclusively observed. Rather it is an indicator of a general form of self-relation that permits a whole series of productive implementations of observation and description to be made concrete. Among these self-relations, a first-person perspective can be differentiated categorically from the perspective of the third person. *My* pain due to an illness, *my* feelings about the results of a genetic investigation, or *my* changed body awareness following knowledge of a genetic risk, are fundamentally distinguished from a description of pain, feelings or perceptions made by a neutral observer. If we speak of the observation from the perspective of the first person, the exclusivity of this perspective is easily marked 'personally' by the use of the pronoun. The concept of person shows here a special perspective that cannot be taken by anyone else. But if we follow Plessner's considerations, the concept of the person is not based only in the otherwise unavailable perspective of an involved participant, but also includes that of the observer of the human body. Both

---

[57] Ibid, 303-312.
[58] Ibid, 383.

perspectives – the subjective and the objective – thus encompass significant aspects of what it means to be a person. Personality is thus – like the concept of the human – not determined merely by a list of characteristics that an entity must have in order to count as a person, but requires a particular phenomenally experienceable self-relation. But at the same time, particular aspects of this personal self-relation must be displayed for special purposes – such as in medical or legal contexts – via such a list of characteristics. *Homo geneticus* is both a biological object seen from a third person perspective and a person with a first person perspective.

Conclusion

While the search for 'genetic transparency' has generated a flow of data on the human genome and genetic diseases, in practice it seems to have generated more uncertainty than solutions. Yet the increasing importance of genetic data for scientific and medical purposes has not led to the hegemony of genetic determinism or to fatalism[59], as many authors initially feared[60]. Despite attempts to ascribe special value to genetic information and thus require a particular 'moral' response from the autonomous and prudent person, personal responsibility can still escape the normative grip of genetics. There is little evidence that human behaviour, health and identity are being reduced to and determined by their genetic basis, not only because of the complexity of genes' functions and interactions with their environment, but also because the subjectification of human being through genetics is only one of various ways of thinking about human individuality in somatic terms[61]. What genetics *means* for being alive and being human is not given by genetic information. Moreover, other forms of expertise coexist and compete with biomedical expertise. *Homo geneticus* remains a cultural concept.

---

[59] Kerr, 2004.
[60] Dreyfuss/Nelkin, 1992; Lippman, 1991.
[61] Rose, 2006.

Literature

Arendt, Hannah. The Human Condition. University of Chicago Press, Chicago, 1958.

Armstrong, David. The rise of surveillance medicine. In: Sociol Health Illn 17(3), 1995, 393-404.

Beauchamp, Tom L. The failure of theories of personhood. In: Standing on principles. Collected Essays, Oxford University Press, Oxford, 2010, 247-260.

Bloom, D.E. et al. The Global Economic Burden of Non-communicable Disease, World Economic Forum, Geneva, 2011.

Bredenoord, Annelien L., Dondorp, W., Pennings, Guido, De Wert, Guido. Ethics of modifying the mitochondrial genome, Journal of Medical Ethics 37 (2011), 97-100.

Bredenoord, Annelien L., Pennings, Guido, De Wert, Guido. Ooplasmic and nuclear transfer to prevent mitochondrial DNA disorders: conceptual and normative issues. In: Human Reproduction Update 14 (2008), 669-678.

Cassell, Eric J. The person in medicine. In: International Journal of Integrated Care 10 (2010), 50-52.

Cassirer, Ernst. Philosophie der symbolischen Formen, Vol. 1: Die Sprache, Meiner, Hamburg, 2010.

Cassirer, Ernst. Der Begriff der symbolischen Form im Aufbau der Geisteswissenschaften. In: Cassirer, Ernst. Gesammelte Werke, Vol. 16: Aufsätze und kleine Schriften (1922–1926), Wissenschaftliche Buchgesellschaft Darmstadt, 2003.

Cassirer, Ernst. An Essay on Man, Yale University Press, New Haven, 1948.

Christman, John. Relational Autonomy, Liberal Individualism, and the Social Constitution of Selves, Philosophical Studies 117(1-2) (2004), 143-164.

Christman, John. The Politics of Persons. Individual Autonomy and Socio-historical Selves, Cambridge University Press, Cambridge, 2011.

Craven, L., Tuppen, H.A., Greggains, G.D. et al., Pronuclear transfer in human embryos to prevent transmission of mitochondrial DNA disease, Nature 465 (2010), 82-85.

Daniels, N. Just Health Care, Cambridge University Press, Cambridge, 1985.

Darwall, Stephen. The second person standpoint. Morality, respect and accountability, Harvard University Press, Cambridge, 2006.

de Waal, Frans. Primates and Philosophers. How Morality Evolved, Princeton University Press, Princeton, 2009.

Dreyer, Malte, Rehmann-Sutter, Christoph, Erdmann, Jeanette. Sequenzen und Menschen. Zur Transparenz genetischer Information, MERKUR, Zeitschrift für europäisches Denken 778(3) (2014), 260-267.

Donagan, Alan. The Theory of Morality, University of Chicago Press, Chicago, 1977.

Dworkin, Gerald. The theory and practice of autonomy. Cambridge University Press, Cambridge, 1988.

Gadamer, Hans Georg et al. (eds.). Neue Anthropologie, Thieme, Stuttgart, 1972.

Gehlen, Arnold. Der Mensch. Seine Natur und seine Stellung in der Welt, Junker und Dünnhaupt, Berlin, 1940.

Foucault, Michel. Les mots et les choses, Gallimard, Paris, 2008.

Harris, J. Could we hold people responsible for their own adverse health? Journal of Contemporary Health Law & Policy 12 (1995), 147-153.

Herbrand, C. Nuclear Transfer Techniques for mitochondrial disorders: How to conceptualise them ethically with respect to the germ-line therapies?. In: Law and the Human Genome Review Special Issue 2014, 2014, 243-249.

Huizinga, Johan. Homo Ludens: Proeve Ener Bepaling Van Het Spelelement Der Cultuur. Wolters-Noordhoff, Groningen, 1938.

Hubbard, Ruth. Predictive genetics and the construction of the healthy ill. In: Suffolk University Law Review 27(4), 1993, 1209-1224.

Lemke, Thomas. Disposition and determinism - genetic diagnostics in risk society. In: Sociological Review 52(4), 2004, 550-566.

Manson, Neil C., O'Neill, Onora. Rethinking Informed Consent in Bioethics, Cambridge University Press, Cambridge, 2007.

Metcalfe, K. et al. Contralateral mastectomy and survival after breast cancer in carriers of BRCA1 and BRCA2 mutations: retrospective analysis, British Medical Journal 2014. doi: http://dx.doi.org/10.1136/bmj.g226.

Miko, Ilona. Phenotype variability: penetrance and expressivity, Nature Education 1 (2008), 137.

Minkler, M. Personal Responsibility for Health? A Review of the Arguments and the Evidence at Century's End, Health Education & Behavior 26 (1999), 121-141.

Morris, Charles W. Foundations of the Theory of Signs, University of Chicago Press, Chicago, 1966.

Munsoin, R., Davis, L.H. Germ-line gene therapy and the medical imperative, Kennedy Institute of Ethics Journal 2 (1992), 137-158.

Naidoo, B. et al. Smoking and public health: a review of reviews of interventions to increase smoking cessation, reduce smoking initiation and prevent further uptake of smoking. 1st Edition, Health Development Agency, NHS, 2004.

Nuffield Council on Bioethics, Novel techniques for the prevention of mitochondrial DNA disorders: an ethical review, Nuffield Council on Bioethics, London, 2012.

Plessner, Helmuth. Die Stufen des Organischen und der Mensch, Suhrkamp, Frankfurt, 1980.

Ritter, Joachim. Subjektivität. Suhrkamp, Frankfurt am Main, 1974.

Rose, N. Personalized Medicine: Promises, Problems and Perils of a New Paradigm for Healthcare, Procedia - Social and Behavioral Sciences 77 (2013), 341-352.

Sommer, Volker, Hof, Jutt. Apes like us. Ed. Panorama, Mannheim, 2010

Steakley, L. What personal DNA testing can reveal about your potential health and future well-being, Scope, Stanford Medicine, 2012. (URL: http://scope blog.stanford.edu/2012/01/03/what-personal-dna-testing-can-reveal-about-your-potential-health-and-future-well-being/#sthash.wlAtVBtr.dpuf).

Stirrat, G.M., Gill, R. Autonomy in medical ethics after O'Neill, Journal of Medical Ethics 31 (2005), 127-130.

Wikler, D. Personal and social responsibility for health, Ethics and International Affairs 16 (2002), 47–55.

Wulf, Christian. Der Mensch und seine Kultur, Anaconda, Cologne, 2011.

4

Personal Genomics: Transparent to Whom?

Teresa Finlay, Shannon Gibson, Lene Koch, Sara Tocchetti

Introduction

Personal genomics is the sequencing and analysis of an individual's DNA to establish knowledge about their genotype. A person's selected single nucleotide polymorphisms (SNPs), exome (protein coding regions of DNA) or whole genome can be analysed and compared to published genomic data for information about common disease risks, specific genetic mutations, genetic disorder carrier status, response to certain drugs (pharmacogenomics), physical traits and ancestry tracing. Over the last decade personal genomics has become available to the public by 'direct-to-consumer' (DTC) marketing and sales and less commonly, by citizen science biohacking activities, known as DIYBio. At first glance these methods for obtaining information about one's genetic makeup may seem like an extension of personal biographical knowledge, rendering one's previously hidden, internal body more transparent. However, the contingent nature of genotyping data means that there is doubt about the utility of the information gained and the benefits of doing so. According to critical commentators and also the federal authorities in the United States (which will be discussed in this chapter), direct-to-consumer genetics testing (DTCGT) lacks clinical validity or utility despite the tropes of personalised health care benefits with which DTCGT companies market their products. Those in favour of DTCGT and DIYbio champion the rights of individuals to access their genetic information, while many ethicists, lawyers and medical professionals think a vulnerable public needs to be protected by regulation.

This chapter considers the issue of transparency in relation to personal genomics from the perspectives of one country's Council of

Ethics' response to DTCGT, the specific issues related to DTC pharmacogenomics and the DIYbio movement. It starts with an exploration of the historical factors that facilitated the emergence of DTCGT, specifically three conditions of possibility necessary for DTCGT to be provided. It should be noted that since the planning of this volume, significant changes in the landscape of personal genomics have influenced the current provision and marketing of DTCGT. There have been takeovers of two major personal genomics companies, DeCODE and Navigenics, with the loss of their SNP genotyping services, and the cessation of marketing of health-related genotyping by 23andMe following a Cease and Desist order by the US Food and Drug Administration in November 2013.[1] Subsequently 23andMe have begun marketing their health-related genotyping product in countries where there are as yet no legal restrictions to prevent them, including Canada and the UK. These changes are indicative of the precarious nature of personalised genomics, which has yet to be negotiated and stabilised as a new genetic technology, and it is with this caveat that this chapter is presented.

Conditions of Possibility for Personal Genomics

There are three conditions of possibility associated with the emergence of DTCGT. These can be grouped into the technological, the ideological and the ethical (or moral order). The concept of genetic transparency is relevant across all three aspects: technological developments have provided significantly greater knowledge and access to that knowledge, rendering it more transparent; ideological changes in society in the late 20th and early 21st centuries have invested the public with more autonomy and raised their expectations of the transparency of their personal genetic information; genetic transparency in relation to moral order poses challenges, as there is tension between individuals' desire for personal information and agency in relation to it, and the historic model of medical paternalism in the provision of healthcare. Each of these conditions of possibility

---

[1] Conley, 2013.

will be examined in turn and the relationship of each to genetic transparency proposed.

Technological Aspects

Developments in both information technology and genomics have provided the technological basis for personal genomics. Since 1991 the World Wide Web has made centrally stored information available to anyone with a computer connected to the Internet. Web 2.0 followed, facilitating interaction and user-generated content to be displayed, which enabled the development of online social networking and commercial activity. As hardware costs have fallen people have increasingly been able to use the Internet for personal, commercial and professional communication, information storage and access, and trade.[2] These developments play a key role in democratisation and personalisation, particularly in relation to Web 2.0, making the Internet the perfect vehicle for DTCGT.[3] Indeed O'Riordan describes the Internet as the "architecture of participation" embraced by digital publics who use social networking technologies.[4]

Most influential for the possibility of DTCGT was the project to sequence the human genome. Begun in 1990, the project was undertaken by the International Human Genome Consortium at university laboratories in six countries, with parallel work being done by Celera Genomics, a biochemical technology company founded by Craig Venter.[5] Completion of a functional 'map' of the human genome was completed earlier than anticipated in 2003[6] and paved the way for genome-wide association studies (GWAS). GWAS were made possible by the work on mapping the human genome and the biotechnological advances in sequencing apparatus. The aim of GWAS is to draw correlations between human genotype variations and diseases by genotyping DNA from large numbers of donors to human biobanks. Genotypes of people with and without the diseases

---

[2] Ward, 2006; O'Reilly, 2009.
[3] Forster/Sharp, 2008; Arribas Ayllon et al., 2001.
[4] O'Riordan, 2010, 38.
[5] Wright et al., 2011.
[6] IHGSC, 2004.

of interest are analysed and compared for nucleotide variations that can be linked to the conditions.[7] The genotypes of SNPs for an increasingly large number of conditions and traits are being established through GWAS. These data provide the template information against which DTCGT companies compare customers' genotypes and calculate disease risk for complex multi-factorial diseases.[8]

There are a number of issues associated with SNP analyses for common complex disease risk evaluation relating to the tests' utility and validity. A genetic test is said to have analytic validity when it accurately detects the genetic anomaly being tested for; clinical validity relates to how well the genetic anomaly indicates presence of disease, and clinical utility indicates the ability of the test to provide information about diagnosis and treatment that is of use to the affected individual.[9] DTCGT is widely reported to have low clinical validity and utility in comparison with predictive genetic testing for single gene defects.[10] Janssens et al.[11] reported the lack of sufficient evidence for SNPs' usefulness as disease risk information, while Ng et al.[12] and a GAO report[13] showed that comparison of test results by some of the larger DTCGT companies yielded different disease risk estimates for the same DNA. In addition, SNP analysis ignores other genetic factors that may affect an individual's propensity to develop disease. SNPs are thought to contribute no more than about 10% to the overall risk of disease; environmental factors are a much more powerful influence on an individual's disease risk.[14] The analytical validity of many companies' SNP analyses has been called into question with the best-known company 23andMe most recently

---

[7] Kaye et al., 2009.

[8] Edleman/Eng, 2009; McBride et al., 2010.

[9] Holtzman, 1999.

[10] Kuehn, 2008; Van Ommen/Cornel, 2009; Patch et al., 2009; Annes et al., 2011; Evans et al., 2011.

[11] Janssens et al., 2008.

[12] Ng et al., 2009.

[13] Kutz, 2010.

[14] Ng et al., 2009.

having to cease trading its health genomics testing from the USA until evidence of validity can be demonstrated.[15]

Pinker and Richards both undermine the deterministic nature and transparency of personal genomics information in their self-reported experiences of DTCGT. They outline the subjective way DTCGT company scientists choose GWAS data for associations between SNPs and phenotypes, and the absurdity of testing for some of the traits included when obvious phenotypes either indicate the same information or contradict it (such as eye colour or ear-wax type). The advice to eat healthily and exercise more applies as the intervention for almost all results but as they note, one does not need to pay for SNP genotyping to know that.[16] However, Pinker also writes about individual curiosity, the entertaining aspects of personal genomics, and the democratic argument for freedom of access to personal information rather than paternalistic regulation, which he and others support.[17] This emerging 'democracy' is key to the ideological conditions for DTCGT.

Ideology

Since the latter half of the 20th century western governments have led the development of neoliberalism, shifting the emphasis from the public to the private, from state controlled public services to business models and private corporations. The developing emphasis on global capitalism has influenced healthcare provision and provided opportunities and expectations for choice and individualised consumption of healthcare services. Devolvement of responsibility for healthcare to individuals places responsibility for health, health promotion and disease risk management on the citizen who has become an autonomous consumer of healthcare services.[18]

Increased autonomy in relation to consumption of healthcare requires individual responsibility and knowledge of health and disease in order for citizens to effectively manage their health and the potential risks to it. This need for information was exploited by

---

[15] Hall et al., 2009.
[16] Pinker, 2009; Richards, 2010.
[17] Pinker, 2009; Vorhaus/MacArthur, 2010.
[18] Arribas Ayllon, 2010; Dickenson, 2013.

personal genomics companies who seized the opportunity for commercial ventures that private consumerism and advancing technology provided. Many companies' branding deliberately emphasised the first person; DeCODE*Me*, Kno*me*, *My*Genome and 23and*Me* (author emphasis), seeking to attract the individual with hyped promises of access to personal health information.[19]

When DTCGT companies appeared in number in the late 2000s, media coverage of DTCGT as an emerging technology often referred to high profile "spit parties" and celebrities' experiences of being tested. With potential customers being attracted to DTCGT with promises of personalised, genomic information that would provide previously unknown information about their health and/or ancestry, celebrity as a commodity was also being used to add to the hype and marketing of this new technology.[20] In her analysis of personalised medicine as compared to the increasingly neglected importance of public health programmes, Donna Dickenson suggests that the emphasis on the personal at the heart of retail genomics, particularly 23andMe, provides purchasers with their own celebrity-aligned experience and a social networking platform on which to share with others.[21] In other words, people who consider buying a test may identify themselves with celebrities who have publicly tested and shared their results through social and broadcast media. With this approach, commercial genomics companies are arguably promoting genomic transparency to and between individuals. However there are two problems here. First, that transparency cannot be reified due to the unstable nature of current genomic knowledge; second, the companies' tropes of personalisation and determinism are contradicted by their suggestions that disease risk can be favourably altered by life-style changes.

In a prophetic publication in the Hastings Centre Report of 1995, Silverman described the expansion of genetic testing from medical laboratories into private enterprise and forecast the proliferation of genetic testing marketed directly to the consumer. He raised concerns about the impact of genetic testing on individuals' behaviour and

---

19 Borry et al., 2010; Dickenson, 2013.

20 Ferris, 2007.

21 Dickenson, 2013.

lifestyle choices and suggested that regulation is the only "moral influence that can be brought to bear"[22]. This will be examined in the next section on Moral Order, the third condition of possibility for DTCGT.

Moral Order

In liberal societies moral order is central to a system of obligation and accountability for productive relationships between individuals and groups in society. Relationships between the public and health care professionals have been based on trust, on a more or less paternalistic basis, but the neoliberal move towards health consumerism threatens this status quo. DTCGT is particularly problematic in this regard because of the tension it sets up between the public (users) and clinicians. Individuals in both groups are accountable to society, healthcare professionals being additionally accountable to patients, employers and their professional bodies. As part of this accountability the public are being encouraged to take responsibility for their health and manage risks to it. DTCGT is sold as facilitating this risk management and is seen as having personal utility for many purchasers.[23] On the other hand the autonomy that 'lay' individuals are exercising in relation to DTCGT challenges clinicians' accountability to them, because many clinicians view the public's choice to buy genotyping as lacking utility, as its information is misleading and potentially harmful.[24] There is potential here for doctors' ignorance and skepticism to undermine the public's confidence in the healthcare establishment, unless healthcare professionals take up Farkas and Holland's call to assist the marketplace and consumers by guiding them through DTCGT with information.[25] The importance of keeping the public's trust and engaging users, DTCGT companies and healthcare professionals in developing new relationships within a more democratic and

---

[22] Silverman, 1995, 17.

[23] Khoury et al. 2009; Bunnik et al., 2011; Tutton/Prainsack, 2011.

[24] Wade/Wilfond, 2006; Wolfberg, 2006; Van Ommen/Cornell, 2008; Kraft/Hunter, 2009.

[25] Farkas/Holland, 2009.

transparent approach to healthcare is highlighted as a more realistic way forward as the genomic era progresses.[26]

However, the ethical problematisation of DTCGT really relates to the lack of transparency about its ontological status. Although it has emerged from research into human genomics, it is difficult to reduce DTCGT simply to the procedure of SNP analysis. The nature of genetic knowledge is wider than its biological basis because its 'bio-power' affects people's lives, their decision-making, their families and their health risk management. In this way, DTCGT has the capacity to cause harm through encouraging anxiety, drastic interventions that aim to manage the risk of disease, or complacency and lack of disease risk management all on the basis of a test that is of doubtful clinical utility.[27] Bunnik et al.[28] highlight that novel data produced by GWAS research both add to and alter previous DTCGT results, so customers may have to change their views about their health in the light of shifting results, and are repeatedly exposed to an emotional rollercoaster each time they are notified about new GWAS associations with their SNPs. Concerns for the relatives who may be affected by genetic testing results, testing the vulnerable or surreptitious testing (testing people without their knowledge, children in particular), are all potential difficulties and can be viewed in juxtaposition to companies promoting DTCGT as fun for the family or the purchasing of testing for others (whatever the motive).[29]

Privacy is generally discussed as a positive attribute, and is related to autonomy that is being exercised in buying a genotype test online. However, breach of confidentiality in relation to individuals' genotyping data is problematic because potentially sensitive information used by third parties for discriminatory reasons may have significantly harmful effects. Despite the enactment of the Genetic Information Non-Discrimination act in the USA in 2008[30], there are concerns about individuals' insurance and employment should DTCGT data fall into the wrong hands. As indicated earlier in the

---

[26] McGowan/Fishman, 2008; Patch et al., 2009.

[27] Wade/Wilfond, 2006; Wolfberg, 2006; HGC, 2007; Katsanis et al., 2008; Wallace, 2008; Patch et al., 2009.

[28] Bunnik et al., 2011.

[29] Borry et al., 2009; Kutz, 2010; Udesky, 2010; Vorhaus, 2011.

[30] Vorhaus, 2010.

discussion of the take-overs of DeCODE genetics and Navigenics, privacy and confidentiality could be threatened when companies go bankrupt and the fate of customers' samples (if kept) and data is uncertain, regardless of the purchasers' country of domicile.[31]

Anthropologist Mary Douglas suggested that shared classifications are central to moral order in human society, and without a shared or commonly accepted understanding, disapproval is expressed and rituals are enacted to restore order.[32] The equivalent in the present context are the debates for and against DTCGT that appear in the popular and professional media, and the numerous calls for oversight of some kind to protect the public from harm. The overall lack of consistent regulation of DTCGT on both sides of the Atlantic relates to the specific challenges that DTCGT presents; globalisation of information and trade enabled by the Internet has vastly widened the scope of access to DTCGT and undermined the power of local legislation to enforce regulatory oversight across borders.[33] The rapid pace of change both of biotechnologies associated with genomics (and thus the products on sale), and the DTCGT commercial landscape, make the moving target of DTCGT difficult to define and regulate. The lack of consensus about what DTCGT constitutes makes oversight difficult. Different groups view it as personal information, an educational or recreational resource, or a clinical diagnostic test, and each would require a different approach in terms of any oversight, which is no doubt why the patchwork approach to regulatory oversight persists.

Having established how DTCGT came into existence and highlighted the ethical challenges it presents, the next section of the chapter discusses some of these challenges further by examining one EU country's regulatory recommendations concerning DTCGT, and the issues they raise.

---

[31] Foster/Sharp, 2008; etc. Group, 2008; Udesky, 2010.
[32] Davis, 2008.
[33] Jordens et al., 2009.

Regulatory Concerns and Personal Genomics: A Case Study

The ethical challenges of creating and acquiring genomic knowledge have been strongly debated for years, and with the advent of advanced GWAS technology, lowered prices for genetic tests and the emergence of an expanded market for private consumption of genetic knowledge, the controversies have exploded. In 1987 the Danish Parliament created a body to oversee and debate the field of ethical aspects of new medical technology, the Danish Council of Ethics (DCE). Over the years DCE has issued reports on a wide range of controversial new technologies, from IVF in the 1990s to stem cells in the 2000s, and in 2012 a report on GWAS and DTCGT. The DCE issued its report on genome testing after a prolonged debate in Danish society, pointing to what the Council considered the main problems surrounding the uses of genome testing (Danish Council of Ethics 2012). The recommendations considered four points: 1) the justification of genome testing; 2) the self-determination of the examined person; 3) counselling and information; and 4) implications for the public healthcare system. This section discusses the DCE's report on DTCGT.

A Critical View of Genomic Testing

The DCE's general view is that personal genome testing is of low utility, low clinical value, and rarely justified. The tests may compromise the right not to know, the right to self-determination and to privacy. The DEC does not recommend prohibiting the selling of personal genotyping through private companies, but finds that appropriate regulation should be introduced. Information and counselling are considered of the utmost importance to ensure informed choice, and expert physicians or geneticists should ideally provide such information to the public. In its report the DCE distinguishes sharply between research (where genomic knowledge ought not to be communicated to the research subject) and therapy (where information should be communicated with great care and include genetic counselling), and proposes very restrictive practices concerning the communication of findings to the patient/research subject, since autonomy and the right not to know may be jeopardised. The use of such tests in the public health system may be controlled by physicians, but the DCE imagines that the availability of DTCGT on the private market may create a wave of worried citizens who may burden the public healthcare system with requests for help to interpret the complex risk information they receive from private companies via the Internet. The DCE concludes that both private and public uses of genomic information should be carried out with great caution.

Limiting Genetic Transparency

The report is characterised by a protective medical viewpoint, which does limit genetic transparency for the consumer – though not for the medical expert. It frames the use of genomic testing as a medical matter, exclusively motivated by medical concerns. Genetic knowledge is first of all motivated by a desire to know about health and disease. The attitude to knowledge creation and transmission is modelled on a framework which is referred to here as classical medical paternalism.

A basic point concerns the uses of testing. The Council proposes that any decision to test should be made in a very careful manner, lest the tests compromise users' rights to self-determination and privacy. The Council fears not only that the public health service system could be burdened by requests from the worried well, but that private insurance companies may also demand knowledge about its customers' genetics and thereby provide a basis for discrimination. Basically, access to genetic knowledge of this kind is considered undesirable, and a policy of limiting access (and by implication transparency) is the result. Among the solutions offered by the Council is a so-called 'careful' use of genome tests, suggesting improved knowledge provision to citizens. The nature of this knowledge however is a censored version of the full information provided by the test results, information which is only considered relevant for the expert advisor.

The report presupposes that the purpose of seeking genomic information is health-related, that the patient is a vulnerable person to be protected by the state not only from voluntarily sought and privately purchased information, but also from the unforeseen implications of her/his own actions. When it comes to genetic health information, the physician is the expert, and the report presupposes the existence of a classic doctor-patient relationship in which the doctor creates or provides medical knowledge about the patient, to be properly interpreted by the expert. Furthermore the ideal system presupposed by the DCE is a Nordic welfare model with a publicly owned and tax-financed healthcare system controlled by physicians. In this system the distinction between research and therapy is clear-cut and possible ethical dilemmas emerging from medical practice should be solved according to ethical guidelines decided by politicians and physicians. The costs of the publicly financed healthcare system should be monitored carefully through the principles of rational healthcare and should always be the result of rational decisions made by responsible clinicians. An uncontrollable wave of clinically unjustified requests would jeopardise this system. The DCE is characterised by the view that free access to services, full knowledge of genetic information and full transparency of the medical and life style-associated implications are problematic.

An Alternative View

The report tries to document its views with reference to scientific studies of the way people receive and handle genomic knowledge, but an alternative view of the practices of DTCGT has been proposed by numerous social scientists including Richard Tutton and Barbara Prainsack.[34] Using as a case study 23andMe's[35] SNP genotyping test for common complex disease risk, carrier status, pharmacogenomics, physical traits and ancestry (as sold until November 2013 when the US FDA ordered the company to cease and desist from selling health-related tests), it can be shown that 23andMe's product was based on a very different model from the one imagined by the DCE. The relationship between knowledge provider and consumer was commercial, but the configuration of this relationship is not as simple as the risky imaginary conjured up by DCE. The information on testing given on the company's webpages was more complex than imagined by the DCE: medical risk information certainly was a major element in the profile presented on the internet, but the company also advertised genealogical information in the form of ancestry-mapping and other information related to social identity.

---

[34] Tutton/Prainsack, 2011.
[35] 23andMe, 2013.

Furthermore, the role of contributor to the scientific development was offered to consumers, under the headline 23andWe, suggesting a less individualised configuration of the consumer, the company and the information. Finally the home page included a recreational/lifestyle aspect of the purchase: it is fun to know one's genes, and one might even buy a test and give it away as a present, "giv[ing] the gift of knowing" to other loved ones, such as relatives. Test results were of course communicated via the Internet, and expertise was distributed between consumers and providers in new ways that break radically with the classic (paternalistic) patient-doctor relationship.

Social relations are configured in a new way. It is difficult to assess whether the 23andMe providers of genomic information have medical training; all we know is that they are professional internet-businesspeople. At the same time however, consumers were not (only) patients but also donors, activists or just curious citizens in search of an original idea for a birthday present! In this new arena the classic medical and research ethical precautions are challenged as direct transfer of genomic data between persons is made possible. The care taken in the medical research arena not to provide allegedly unethical incentives to become research subjects is replaced by direct encouragement to participate in research, while charging clients a fee for doing so. In this context, information and full transparency are worthwhile and empowering objectives, not dangerous risky activities with negative social implications.

Two Models for Handling Genomic Information

Comparing DCE and 23andMe, we encounter two very different, almost antagonistic models for handling genomic information. One is the classic medical framework idealising medical authority, with the doctor as expert and the patient as passive recipient of processed, medically authorised knowledge. The counter image to this system, as presented by DCE, is the market place where vulnerable citizens are potentially exploited and abused by a cruel, profit-hungry private companies selling uncertain risk information. This counter image however is contradicted by the complex reality evident on webpages like that of 23andMe. The following quotes illustrate how 23andMe actually bridged some of the dualisms that are drawn up in the DCE report, and offered its customers a much more nuanced engagement with genomic knowledge (prior to the cease and desist order served on the company in November 2013). They acknowledged central values also highlighted in the DCE report such as responsibility towards self and others, altruism and communal activity, and of course self-determination and protection of one's own interests.

Responsibility: "23andMe isn't just about you. Our research arm, 23andWe, gives customers the opportunity to leverage their data by contributing it to studies of genetics. With enough data, we believe 23andWe can produce revolutionary findings that will benefit us all."

Communality: "Get involved in a new way of doing research. Direct research by participating in studies of conditions and traits you care about. Join an effort to translate basic research into improved health care for everyone."

Individuality: "Support 23andMe's efforts to discover new genetic associations that could shed more light on your data; participate in research while exploring your own genetics; take surveys that collect important data for scientific research. Learn new things about yourself – and what your genes may have to do with them. Find out which traits make you stand out from the crowd."[36]

Such quotes illustrate the very complex conceptual universe characteristic of the 23andMe homepage. Classic dualisms are undermined and new categories emerge. It is not obvious whether the people buying these services are to be seen as patients or consumers, as patients or research subjects. Neither is it clear whether the providers are doctors or businesspeople; neither or both?

The classic distinction between market and public health service, between tax-financed states or privately funded companies is also challenged. The lay-expert distinction is undermined, the distinction between research and therapy is dissolved; patients and doctors, consumers and providers are all entrepreneurial actors in the marketplace. It seems that there are now two ways of looking at genetic technology. One finds that new technology is potentially

---

[36] 23andMe, 2013.

threatening (for example to individual rights and autonomy, and to a just and rational spending of taxpayers' money). The other finds new technology enabling (for example, by creating new social relations characterised by responsibility and altruism). And whereas the first finds the transmission of genetic transparency to be dangerous, the other sees it as an enabling kind of knowledge to be used for multiple worthwhile purposes.

Understanding the Difference

Taking a step back to consider understanding about genes and the role they play in human life, the DCE's view has its origins in the worried 1990s. This was a time when the fear of eugenics, of state and corporate abuse of individual's genetic information guided the legislative pattern of solving the problems emerging in the wake of the genome mapping projects. This view stated that each individual is genetically unique, and the government of genetic knowledge should first of all protect individual privacy. This appeal to individual interests and idealisation of individual autonomy was promoted to counter the fears of a new eugenics, and indeed fear seems to be the dominant feeling behind the DCE report; fear that individuals might be abused by interests other than those pertaining to their own private self. The 23andMe view, however, not disregarding the commercial and profit oriented motives behind such web-based businesses, implies a more optimistic view of the uses of genetic knowledge, and includes a certain amount of altruistic sharing of knowledge. And indeed, the view of Mendelian genetics, which initiated the age of modern eugenics, was characterised by an understanding of genes as belonging to the social collective, as endowing qualities that are socially relevant, and should be governed by social means in the interest of society. What makes the 23andMe-vision so attractive to modern citizens may very well be its appeal as a member of a collective, that it offers a possibility to people of doing good for others and thus represents a convergence of old individualistic and new collective norms. Barbara Prainsack and colleagues noted: "For the first time we see research participants paying to be enrolled in research projects."[37] This surprising aspect is worth closer study, and the wish to do so cannot just be dismissed as being due to naivety or misunderstanding. This may be seen as a new kind of subjectivisation, a new way of shaping subjects, a new kind of bio-citizenship, in which roles are converging. The customer in the new web-based genomic knowledge shop is patient, donor, expert, activist and consumer at the same time; she is seeking knowledge as well as entertainment, and takes responsibility for her own health as well as that of others. Her interests include medicine, lifestyle, research and genealogy, and the compartmentalisation of medicine and other aspects of life do not necessarily make sense. These new subjects may be seen as both enterprising AND altruistic selves.[38]

The Social and Organisational Level

On a more social and organisational level, we may be moving from a public tax-financed health sector in Europe to a private market for genetic tests, a development that holds something of a paradox. In the publicly financed health service sector in countries like Denmark, genetic tests such as those sold by 23andMe are considered problematic by the medical profession and will only reluctantly be offered through the public health sector, if at all. What has happened is not that genomic testing has been halted, but rather that private companies promising individual AND social benefits now market them cleverly and aggressively. The initial success of 23andMe may be seen as a shift in the power of the actors in the genomic marketplace. Patients are no longer docile actors who automatically trust their doctor, while doctors have become authoritative managers who lack their previously unchallenged power. Patient advocacy groups challenge medical norms of regulation (for example medical indication versus personal choice), and the focus and conduct of research, the ownership and benefit of research results. The DCE view of personalised genomics illustrates a conflict between governing citizens and genetic knowledge, a protective paternalism resting on medical authority and a liberal empowerment resting on

---

[37] Prainsack et al., 2008, 34.
[38] Tutton/Prainsack, 2011.

individual choice. Both idealise[39] 'good' governance that derives from decisions that are uninfluenced by political and economic forces. One idealises medical advice and focuses on expertise and authority, the other idealises individual choice and focuses on issues of control. However, both have limitations, particularly in relation to transparency, and future debates on the government of genomic knowledge must take empirical knowledge about users and the market into consideration.

The following section of this chapter focuses on a specific aspect of personal genomics: pharmacogenomics. This aspect of genomics has attracted much hype about the expectations of personalised medicine, most of which have yet to materialise.

Direct-to-Consumer Pharmacogenomic Testing

Pharmacogenomics is the study of the role that genetic factors play in people's responses to drugs. The term "pharmacogenomics" derives from the fact that "this new field combines the science of how drugs work, called pharmacology, with the science of the human genome, called genomics."[40] Scientists have known for years that genetic traits can have a significant impact on the safety and efficacy of a drug for a particular patient. Depending on each individual's genetic makeup, some drugs may work more or less effectively, or may produce more or fewer side effects. However, recent scientific advances have improved understanding of how genes and genetic variation contribute to drug response.[41] In time, pharmacogenomics may enable doctors to routinely use an individual's[42] genetic information to select which drugs and drug doses are most likely to benefit a given patient.[43]

As noted by Kalf et al.,

---

[39] Rose/Novas, 2004.
[40] NHGRI, 2014.
[41] Kerr et al., 2011.
[42] Kerr et al., 2011.
[43] NHGRI, 2014.

> "[t]o date, most of the discussion about DTC personal genome tests has focused on the prediction of these multifactorial diseases and less attention has been given to the predictive ability and clinical utility of pharmacogenetic testing."[44]

Nonetheless, the few major players in the DTCGT market, including 23andMe, have been offering results on gene-drug associations for years, as have smaller players such as GenePlanet and Theranostics. One company, YouScript, even focuses specifically on pharmacogenomic testing: the company bills itself as

> "a revolutionary way of using genetics and the latest clinical knowledge on drug metabolism to help predict which prescription medications will work best for a person."[45]

23andMe reported that one of the goals of its research is to discover novel pharmacogenomic associations using web-based phenotyping of efficacy and toxicity.[46] Moreover, Anne Wojcicki, a co-founder of 23andMe, has stated that one of the driving forces behind the establishment of the company was the potential to test her child's genome before administering medication.[47] When they were marketing health-related testing from the USA, the health overview provided to consumers by 23andMe included a summary of results for diseases for which a person may be at greater than average genetic risk, heritable diseases for which a person carries one or more genetic variants (carrier status), and drugs for which a person is likely to have an atypical response based on genetics.[48] 23andMe's customised genotyping chip tested numerous SNPs known to be associated with drug metabolism, efficacy, toxicity, or other side effects as stated in a post on the company's[49] blog.

As with other areas of DTCGT, there is significant concern about the clinical validity and utility of the testing offered by many

---

[44] Kalf et al., 2013, 341.
[45] YouScript, 2014.
[46] News Medical, 2010.
[47] Shiels, 2008.
[48] Prainsack/Wolinsky, 2010.
[49] 23andMe, 2010.

companies. Consequently, there is widespread concern that many of the clinical benefits of pharmacogenomics and personalised medicine have been oversold and there is a risk of overstating the value of pharmacogenetic testing. As stated by Prainsack and Wolinksy, "one should not forget that those who have a stake in selling a product are usually not entirely unbiased in describing its value and importance."[50] Other critics object to DTCGT because it reduces the role of health professionals as gatekeepers to complex genetic information.[51]

Similarly, the most significant concern with DTC pharmacogenomics is a lack of evidence of the clinical value of many of the tests on offer: "Two primary problems are the quality of evidence that links genetic variants to functional effects, and the clinical utility of genotypes for specific genes."[52] For example, in a 2012 study comparing pharmacogenomic testing offered by eight DTC companies, Chua and Kennedy showed that genetic testing was clearly recommended by the US Food and Drug Administration for only four out of 30 drugs reviewed. Genetic testing for many of the remaining drugs may not have been approved due to a lack of empirical studies demonstrating their predictive ability.[53]

Germino and Chan argue that DTCGT for drug-gene associations in pharmacogenomics may have greater clinical utility than other types of DTCGT. Since certain genes or gene mutations can significantly increase the risk of adverse events from specific drugs, they argue that with pharmacogenomics testing, "[i]ndividuals would be able to take an alternative drug from the onset, reducing their risk of life-threatening complications and saving time and resources."[54]

The samples and data collected by DTC testing companies may also serve as an important platform for studying gene-drug interaction. For example, in 2010, 23andMe received funding from the US National Institutes of Health to investigate genetic factors underlying responses to three classes of drugs: non-steroidal anti-

---

[50] Prainsack/Wolinsky, 2010, 652.
[51] Lenzer/Brownlee, 2008.
[52] Chua/Kennedy, 2012, 7.
[53] Chua/Kennedy, 2012.
[54] Germino/Chan, 2013, 281.

inflammatory drugs; proton-pump inhibitors used to treat gastroesophageal reflux disease; and the blood thinner, Warfarin.[55]

Since it will likely be many years before health systems begin to adopt routine, widespread pharmacogenomic testing, Chua and Kennedy predict that DTC pharmacogenomics companies are well positioned to dominate the market in the near future.[56] However this confidence has been rendered uncertain by recent developments in the DTCGT market, as described at the beginning of the chapter. Flying in the face of large genomics companies and 'big pharma'[57] is the DIYbio network whose activities enable an alternative take on genetic transparency to interested citizen scientists. Their activities are outlined and discussed next, in the final section of this chapter.

Do-It-Yourself Genotyping: Genetic Transparency Gets Hands-On?

This section explores a specific ramification of personal genomics named 'Do-it-yourself genotyping' (DIY genotyping), a term used here to name transient practices of DTCGT established by members of the Do-it-Yourself Biology (DIYbio) network. Following Haraway, who argues that "transparency is a peculiar form of modesty"[58], this section aims to demonstrate that DIY genotyping practices are useful for thinking about genetic transparency. This is because they are located at the intersection of the politics of vision of modern sciences and contemporary technosciences, and the politics of information freedom as informing the latest recursive implosion of biology (in this case genetics) and informatics. More specifically, the aim here is to contribute to the following arguments: the element of interactivity that genomes gain when being[59] incorporated[60], the figure of the learned person as a specific type of subjectivity hailed

---

[55] News Medical, 2010.

[56] Chua/Kennedy, 2012.

[57] Chua/Kennedy, 2012.

[58] Haraway, 1997, 26.

[59] Delfanti, 2013; Doyle, 1997; Fox Keller, 1995; –, 1996; –, 2002; Haraway, 1997; Landecker, 2007; Oyama, 2000.

[60] O'Riordan, 2010.

into existence by personal genomics[61], and the critique of the participatory turn in personal genomics[62]. The main questions this contribution therefore addresses are: what is the socio-technical promise of DIY genomics and its mode of address? Who is the public of DIY genomics? And what is the participatory promise of DIY genotyping? Four examples of DIY genotyping will be presented, based on fieldwork[63], followed by an analysis of DIY genomics in relation to the arguments of O'Riordan, Reardon and Prainsack cited above.

A Brief Portrait of the DIYbio Network

The first meeting of the DIYbio network was held in a pub near the Massachusetts Institute of Technology (MIT) in May 2008. The network has since developed local groups and community laboratories in a number of centres characterised by fast-paced entrepreneurial, techno-scientific and cultural capital such as New York, the Bay Area, Los Angeles, Paris, Manchester, London and Copenhagen. Despite the network's geographical extent the number of active members is estimated at less than fifty. The socio-technical vision inspiring the members of the DIYbio network is highly composite. It draws on the craft movement, computer hackers and open science ethics, and citizen science.[64] It also draws on the socio-technical vision of the personal computer as a central technology in a revolution that would bring to life an ideal society, decentralised, egalitarian, and free. The practices in which DIYbio members engage include the fabrication of affordable, user-friendly and mobile laboratory instruments that are used in combination with purchased materials and educational kits to carry out hands-on activities. These include the extraction of DNA from fruits or buccal scrub samples, the identification of species or phenotype distributions by amplification of genetic polymorphisms, the growth of bacterial and

61 Reardon, 2010.
62 Prainsack, 2011.
63 O'Riodan, 2010; Reardon, 2010.
64 Delfanti, 2013; Roosth, 2010; Meyer, 2012.

fungal biomaterials, the preparation of various fermented products, bacterial transformations, and DIY genotyping experiments.[65]

DIY genotyping: hands-on interactivity, entertaining genomes and extreme participation?

In 2009 Kay Aull, a recent graduate from the Biological Engineering program at the Massachusetts Institute of Technology and an early member of the DIYbio network, became known for performing a genetic test on herself. In our interview Aull mentioned that having a laboratory in her closet was not unusual for her. During her childhood she often carried simple biology experiments and participated in science Olympiads. In 2008 she joined the "Mad Science Contest"[66] organised by a science and science fiction group. Aull presented a project aimed at genetically engineering cells to make them count in binary numbers. This project attracted the attention of journalists and the curious lay public alike, but she recalled experiencing repeated difficulties explaining its purpose. For these reasons she designed a different one. She explained:

> I wanted a project that would be for the people just coming in [amateur biology], have people's attention, I wanted to put together a project that could be done with very basic equipment that you would have wanted in a wet lab and that it would be appealing to a big range of people.

A few years earlier, her father had a genetic test for haemochromatosis, a hereditary and potentially lethal genetic condition, and was found to be positive. Aull remembered her mother complaining. She could not read the results because they were "written in genetics." On the contrary, her father "the engineer" as she called him, started "googling all the stuff he could not understand". This convinced her of how relevant it was to bring genetic tests "from the category of I can't understand it, to the one of I don't [understand it] but I can find out". She therefore designed an experiment aimed at demonstrating that "you could actually do it. It's

---

[65] Tocchetti, 2012.
[66] Roosth, 2010; Whalen, 2009.

not commonly realised that you can just look inside your genome, it's kind of a mysterious thing that you have to get a doctor to help you with". She presented her preliminary results at CodeCon, a renowned computer hacker's conference. During her presentation, entitled "Homebrew Genetic Testing" she reproduced part of the procedure, passing the bag with the agarose gel on which she had run her PCR-amplified DNA samples around the audience. As if back there, she said, "you see that little tiny smudge? Maybe that's a mutation that could kill you if you're not careful with it. Just having a physical artefact made it real for people".

Her experiment got a lot of attention from the press, to the point that a journalist filmed her reproducing the experiment in her flat and uploaded the video online.

Shortly after CodeCon, Aull participated in another DIY genotyping experiment, this time a public one. Together with Mackenzie Cowell, the co-founder of the DIYbio network and other members of DIYbio in Boston, they proposed that attendees at the 2009 Cambridge Science Festival should "Test Your Genes". One of the genotyping tests they proposed to the attendees was for BRCA1, a gene implicated in breast cancer. Aull added that because the sequence was still patented by Myriad Genomics at that time, "we kind of wanted to do a civil disobedience project". Unfortunately, the patent claim was invalidated one week before the science festival, taking the wind out of their 'rebellious' gesture. She also mentioned that participants who did not want to know about their BRCA1 genotype could choose a recreational genetic test such as the "bitter-tasting gene." Being the only female at the table, Aull also joked about having to "do the genetic counselling", and sarcastically added that her male colleagues had nothing to feel smart about as BRCA1 is also a candidate gene for prostate cancer.

While Aull's examples highlight the importance that DIYbio members give to physical artefacts reifying the promise of looking into one's genome, other members dedicate themselves to the design, fabrication and distribution of basic, affordable and easy-to-use laboratory hardware. One of the most iconic cases is that of Tito Jankowski, Josh Perfetto and Norman Wang. These three early members of DIYbio designed an Open Gel Box and Pearl

OpenPCR, respectively an "open-source and hackable"[67] electrophoresis chamber and "a low cost, accurate thermocycler you build yourself"[68]. In 2010, at the San Francisco Maker Faire, an event refereed at as the "Woodstock of DIY"[69], they intercepted visitors asking, "Hate Brussels sprouts?" and offered experience of genotyping "the responsible gene". In a video recorder at the stand (and uploaded on YouTube), Jankowski explained how both the Open Gel Box and the OpenPCR could be used to "look at our own DNA and figure out what our DNA says about that bitter taste capability". After showcasing the tools and the process, Jankowski attempted to persuade the viewer that:

> It's really a simple analysis; either you do or you don't and we can show you all the techniques and the cool things. You don't have just to look at the bitter-tasting gene, you can take this technique and look at anything in your DNA. This is one letter out of three billion letters in your genetic code there are three billions of other things that you can do!

Jankowski and Perfetto's DIY genotyping experiment is similar to the one proposed by Aull. But if their proposition reiterates part of Aull's rhetoric and practices, they additionally use DIY genotyping experiments to promote the dissemination of open-source laboratory hardware, which in return is used during events promoting the dissemination of DIY genotyping. For instance Perfetto showcased and used the OpenPCR during a workshop entitled "Sequence your DNA" that he led at the Hacker Dojo in Mountainview, California in May 2011. During the workshop Perfetto compared his proposition to the service available from the company 23andMe. He argued that he could not compete with their service in terms of the number of SNPs screened. Instead, he explained that the workshop was more engaging and participative as it allowed attendees to do all the steps by themselves, except for the sequencing, which he described as the "not exciting" part.

---

[67] Delfanti, 2013; Delgado, 2013; Meyer, 2012.

[68] Pearl Biotech, 2014; OpenPCR, 2013.

[69] Metzger, 2010.

The final example is from an event co-organised by members of MadLab and The Arts Catalyst called LabEasy. This two-week residency was co-organised by the Arts Catalyst, a London-based arts organisation that experimentally and critically engages with science, and members of MadLab, the Manchester Digital Laboratory for hackers and craftspeople. For the occasion several international members of the DIYbio network were invited to further "explore Do-It-Yourself biology," while offering a variety of workshops and participatory events open to "all". One workshop in the residency was headlined "Cocktails and Canapés – an evening exploring genetics and flavour". This was a DIY genotyping experiment targeting the "bitter-tasting gene" (the same as that of Jankowski and Perfetto at the Maker Faire). While the atmosphere was similar to the other workshops I attended, playful, informal and loaded with excitement, on this occasion the ambience was enhanced with soft, coloured lighting and music, similar to a lounge bar. This event was the first of its type to require consent. The short 'consent' document stated:

> You don't have to take part in this part of the evening [the genetic test], but if you do:
>
> You are giving us your DNA! We could do all kinds of things with it, like determine if you are made from horse meat or are a genetically modified vegetable. We could also splash it around the scene of a hideous crime. However, we promise not to: we will keep it safe (in the freezer in the corner) and use it instead to search for answers to intriguing cocktails and genetics-based questions.
>
> In exchange for your DNA we will give you a number. You are no longer a name but a number. When we release the results, we'll number them so you can see which is yours. This isn't 100% anonymous obviously and you might want to think about that. Tell your friends!
>
> There will be spurious graphs.
>
> We get things wrong. We like to think that what we tell you will be correct, but we can't guarantee it. Hell, this whole experiment might not even work at all. Anyway, understand this and if you have any concerns or questions

> about your palate as a result of our experiment, consult a chef.

Below these four points, the participants wrote their names and signatures. Judging by conversations at the welcome table, it appeared that none of the participants questioned the necessity of the form, refused to sign, or openly commented on the style of the document. From the beginning the tone of the document suggested a casual, amused tone, with a hint of reflexivity. The organisers invoked the participants' trust by joking about far-fetched ways in which they could misuse their DNA samples, the first of which does not even make sense. The second point is marked by an invitation to think about the issue of anonymity in respect of gene sequences. The fourth and last point is again an amusing warning concerning the reliability of the experiment. In line with its beginning, the document ended with an ironic joke.

DIY Practices and Personal Genomics

These are sporadic practices and preliminary observations. But it is arguable that as the subjectivities of DIYbio members are hailed into existence by the genotyping practices characterising personal genomics (and specifically DTCGT), DIY genotyping will become a point of circulation and a site of their reconfiguration. What circulates is the socio-technical vision of the genome as a hermeneutic site whose rights of access are defended from doctors and other gatekeepers, in the name of the individual's right to know. With DIY genotyping, looking into one's own genome is confirmed as a personal service but is then augmented into becoming a personal right. However, what distinguishes these practices is the promise of 'hands-on' interaction with one's own genome.[70] As Aull explained it is seeing the material artefact that allows participants to understand the meaning of a genetic mutation. Similarly Perfetto argues that while his workshop cannot compete with 23andMe, it offers a more exciting and interactive experience.

As part of recent debates on human genomes and media cultures, O'Riordan's analysis of personal genomics proposes that a shift from

---

[70] Rainbow, 1997.

iconography to incorporation is on going. While human genomics was characterised by the metaphor of[71] "the book of life" and audiences marvelling at the iconography of the gene, "the current address invites audiences to come closer and to interact with genomics, incorporating it into embodied practices"[72]. If O'Riordan mostly refers to initiatives such as 23andMe's SNP genotyping and other commercial personal genomics services, DIY genotyping illustrates that "hands-on" is an additional form of interactivity in-the-making, a material-semiotic mode of address that is informed by reconfigured laboratory and DIY practices.

These practices are presented as interactive, transgressive (as in the case of the "civil disobedience" initiative), but also highly entertaining. Jankowski promotes DIY genomics as the ultimate antidote to boredom, and suggests that hands-on interaction with one's own genome is as much about a relationship between genes and health as it is a recreational activity promising fun and entertainment. As part of Jankowski's proposition, the rhetoric of the endless possibilities of genomics becomes an entertainment value accessible to a larger public. This means that DIYbio members also use DIY genotyping as a means to address specific types of publics as part of their community-building effort, a fundamental gesture for a group whose core socio-technical promise is that of biology and biotechnology being available to all. Science festival attendees are known to enjoy following science and technology developments. By being showcased on the Internet, during science festivals, Maker Faires and arts/science events, the type of publics known to characterise DTCGT attract new members. More specifically, the demography of the "maker movement" is mostly composed of educated or highly educated men, a majority of whom are trained in engineering and science and who practise DIY craft and hacking activities in the company of their families. Similarly, a survey conducted by the organisers of EasyLab indicates that a majority of the participants' professional status was as designers, multi-media artists, science museum employees, science communication professionals, scientists and university students. Therefore these

---

[71] O'Riordan, 2010.
[72] O'Riordan, 2010, 8.

additional members do not represent a rupture within the publics of DTCGT, rather the opposite. In her work on the company 23andMe, Reardon argues that the learned person is a type of subject hailed into existence by the shift of human genomic research from the "vulnerable" isolated indigenous population to the "empowered" person.[73] She highlights how personal genomics companies, in resonance with critiques by science and society scholars, have reconfigured the deficit model of science and society to think and promote their relations to the public. Reardon critiques the composition of this "public" based on the distinction between "in-the-know digerati on the one hand, and all people on the other".[74] The examples presented suggest both that the members of the DIYbio network involved in these activities, and the members of the public addressed by these activities, are in fact part of an elite resembling Reardon's figure of the "learned person"[75].

The participatory models promoted by DIY genotyping may be examined in relation to Prainsack's models of participation in DTCGT: the Napster, the Supermarket and the IKEA models. Each model illustrates several elements of the discussion about what she refers to as the "participatory turn" in personal genomics, the IKEA model being arguably the most apposite to DIYbio. Prainsack aligns 23andMe's[76] invitation to customers, to participate and exercise choice by completing the company's questionnaires and purchasing tailored services, with buying IKEA furniture. "Customers were given the feeling that they were building something themselves" [but eventually, they end up choosing] "from a predetermined range of elements" [and assembling them]. The argument that "building from scratch"[77] is a more participatory model suggests that the issue is not the politics of what constitutes a genetic test, but how the process of carrying it out can be considered more or less participatory. The threshold of participation is therefore 'displaced after' the factual existence of the genetic test.

---

[73] Schiele/Jacobi, 1993; Tocchetti, 2012.

[74] Reardon, 2011, 97.

[75] Ibid.

[76] Prainsack, 2011, 140.

[77] Ibid.

The participatory promise of DIY genotyping was highlighted by the counterintuitive presence of a document resembling a consent form. As in Aull's experience, when she claimed she improvised being an informal genetic counsellor, its presence suggests that the promise of a DIY biology might not be as straightforward as it initially appears. In practice, conventional forms of expertise (including doctors or DTCGT companies) are supplemented by others, in this case DIYbio members themselves who become a type of self-declared non-expert. Similarly, the document's mode of address relies upon a rather sophisticated irony, recalling recent incidents that threatened public trust (in the UK) such as the horsemeat scandal or labelling of food containing genetically modified organisms. The consent document's content[78] suggests that implicit knowledge is required to understand such ironies, a characteristic that contributes to the seemingly sophisticated subjectivity of the learned person. For instance, while the consent form suggests two reflexive opportunities, one concerning the issue of anonymity and the other the possibility that the experiment might simply fail, neither was later addressed by the organisers during the event. Similarly none of the participants questioned the politics of preventative medicine nor, in the case of recreational genetic tests, the type of relationship to oneself that this information promotes. Lastly, the participants were in a specifically reflexive environment. They had the choice between figuring it out for themselves, or asking questions of the organisers, who in the case of the event LabEasy were not geneticists or persons who had acquired a substantial understanding of the socio-technical controversies concerning personal genomics. Yet these examples suggest that this might not be problematic, as the majority of the participants are 'learned persons'.[79]

While Reardon argues that personal genomics has created a powerful zone of biosocial formation by "yoking the locus of agency in liberal democracy – the 'person' – to the locus of agency in the life sciences – genomes"[80], DIY genotyping reconfigures such powerful

---

[78] Lawrence, 2013.
[79] Reardon, 2011.
[80] Reardon, 2011, 95.

biosocial formation by articulating the politics of the person to the powerful social-economy of DIY practices, a locus where the figure of the liberal individual is recursively reconfigured and regenerated. The critique of the liberal individual is common in the study of personal genomics, and is linked to the significant de-regulation of health services that the US and Europe are currently undergoing. Nonetheless it would be premature to argue that the hands-on elements characterising DIY genotyping are simply another locus of powerful biosocial formation in the sense dictated by liberal democracies. As Goldstein has illustrated for the history of the DIY movement, these preliminary case studies call for a suspicious yet engaged analysis of DIY genotyping.[81]

Conclusion

Despite the recent fluctuations in availability of commercial personal genomics to the public, genotyping is an increasingly commonplace technology, whether accessed by individuals curious to make use of the supposed transparency with which this technology has been invested, or whether part of biomedical research or diagnosis. This chapter has illustrated the circumstances that enabled genotyping to be made directly available to the public, and presented three different aspects of personal genomics to illustrate the uncertain and contested nature of DTC genotyping.

As the case studies have demonstrated, there are a number of binary oppositions at work in genetic transparency. There are arguments about who should have access to genetic information: many lawyers, bioethicists and healthcare professionals use a deficit model of public understanding of genomics to suggest that the public would be harmed by having direct access to their genetic information. Conversely some scientists and members of the public feel that this form of transparency is an individual right, and so engage in DIYbio practices or purchase tests from commercial genomics companies in order to exercise their individual autonomy and reject the paternalistic protections of the state mediated through healthcare

---

[81] Goldstein, 1998; Petersen, 2011; Prainsack, 2011.

providers. Indeed it is at times precisely to avoid scrutiny of their genome by healthcare professionals, and by extension insurers or employers, that people purchase DTCGT, and a number of research studies have established that people do not appear to suffer harm as a result. Companies are deterministic in their sales rhetoric but scientists and healthcare professionals repeatedly criticise this framing of genotyping, referring to the flawed nature of GWAS in particular and the lack of understanding of human genomics more generally.[82]

Current knowledge about the human genome and its relationship to health and disease are not at a point where any degree of personal transparency is possible, at least with the personal genotyping practices currently available. Unless and until that point arrives and the knowledge, understanding and application of the technology can be stabilised by negotiation between all the interested parties identified here, personal genomics is likely to remain contested and obscure.

Literature

23andMe. Health at 23andMe: when one size doesn't fit all. 23andMe (2010) (URL: http://blog.23andme.com/23andme-and-you/23andme-how-to/health-at-23andme-when-one-size-doesn%E2%80%99t-fit-all/). Accessed 25 March 2014.

23andMe, research web page; 23andMe (2013) (URL: https://www.23andme.com/research/). Accessed 3 February 2013.

Annes, J.P., Giovanni, M.A., Murray, M.F. Risks of pre-symptomatic direct-to-consumer genetic testing, New England Journal of Medicine 363(12) (2011), 1100-1101.

Arribas-Ayllon, M., Sarangi, S., Clarke, A. Genetic Testing: accounts of autonomy, responsibility and blame, Routledge, Abingdon, 2011.

Bloss, C.S., Schork, N., Topol, E.J. Effect of direct-to-consumer genomewide profiling to assess disease risk, New England Journal of Medicine 364(6) (2011), 524-534.

Borry, P., Cornel, M.C., Howard, H.C. Where are you going, where have you been? A recent history of the direct-to-consumer genetic testing market, Journal of Community Genetics 1(3) (2010), 101-106.

[82] Bloss et al., 2011; Collins et al., 2011; McGowan et al., 2010.

Borry, P., Howard H.C., Senecal, K., Avard, D. Direct-to-consumer genome scanning services: also for children?, Nature Review Genetics 10(1) (2009), 8.

Bunnik, E.M., Schermer, M.H.N., Janssens, A.C.J.W. Personal genome testing: test characteristics to clarify the discourse on ethical, legal and societal issues, BiomedCentral Medical Ethics 12(11) (2011).

Caulfield, T.A., Ries, N.M., Ray, P.N., Shuman, C., Wilson, B. Direct-to-consumer genetic testing: good, bad or benign?, Clinical Genetics 77 (2009), 101-105.

Chua, E.W., Kennedy, M.A. Current state and future prospects of direct-to-consumer pharmacogenetics, Frontiers in Pharmacology 3 (2012), 1-8.

Collins, R.E., Wright, A.J., Marteau, T.M. Impact of communicating personalized risk information on the perceived control over the risk: a systematic review. Genetics in Medicine 13(4) (2011), 273-277.

Conley, J. If 23andMe falls in the forest, and there's no-one there…, Genomics Law Report (2013) (URL: http://www.genomicslawreport.com/index.php/2013/12/03/if-23andme-falls-in-the-forest-and-theres-no-onethere/#more-13198). Accessed 8 May 2014.

Davis, J.E. Moral order. Culture. (2008) (URL: www.iasc-culture.org/eNews/2008.../LastWord_CultureSpring08.pdf). Accessed 8 May 2014.

Delfanti, A. Biohackers: the Politics of Open Science, Pluto Press, London, 2013.

Delgado, A. DIYbio: making things and making futures, Futures 48 (2013), 65-73.

Dickenson, D. Me Medicine vs. We Medicine: Reclaiming Biotechnology for the Common Good, Columbia University Press, New York, 2013.

Doyle, R. On Beyond Living: Rhetorical Transformations of the Life Science, Stanford University Press, Stanford, 1997.

Edleman, E., Eng, C. A practical guide to interpretation and clinical application of personal genomic screening, British Medical Journal 339 (2009), 1136-1140.

etc Group. Direct-to-consumer DNA testing and the myth of personalised medicine: spit kits, SNP chips and human genomics, Special Report on Human Genomics I. Erosion, Technology and Concentration Group (2008) (URL: http://www.etcgroup.org/node/675) Accessed 8 May 2014.

Evans, J.P., Dale, D.C., Fomous, C. Preparing for a consumer-driven age. New England Journal of Medicine 363(12) (2011), 1099-1103.

Farkas, D.H., Holland, C.A. Direct-to-consumer genetic testing: two sides of the coin, Journal of Molecular Diagnostics 11(4) (2009), 263-265.

Ferris, K.O. The sociology of celebrity, Sociology Compass 1(1) (2007), 371-384.

Foster, MW, Sharp, RR. Ethics watch: out of sequence: how consumer genomics could displace clinical genetics, Nature Reviews Genetics 9.419 (2008) (URL: http://ezproxy.ouls.ox.ac.uk:2346/ nrg/journal/v9/n6/full/nrg2374.html). Accessed 8 May 2014.

Fox Keller, E. Refiguring Life: Metaphors of Twentieth Century Biology, Colombia University Press, New York, 1995.

Fox Keller, E. The biological gaze. In: Robertson, G., Mash, M.L., Tickner, J., Bird, B., Putnam, T. (eds.) FutureNatural, Routledge, London, 1996.

Fox Keller, E. Making Sense of Life: Explaining Biological Development with Models, Metaphors, and Machines, Harvard University Press, Boston, MA, 2002.

Germino, N., Chan, K. Simulation of consumer trends in direct-to-consumer (DTC) genetic testing, Journal of Consumer Health on the Internet, 17(3) (2013), 272-283.

Goldstein, M.C. Do-it-yourself: Home Improvements in 20th Century America, Princeton Architectural Press, New York, 1998.

Hall, A., Luheshi, L., Kroese, M., Brice, P., Zimmern, R.. Regulating genetic tests: lessons from 23andMe PHG Foundation December (2013). (URL: http://www.phgfoundation.org/news/15115/). Accessed 8 May 2014.

Haraway, D.J.. Modest_Witness@Second_Millenium.FemaleMan©_Meets_OncoMouse®, Routledge, New York, 1997.

HGC. More Genes Direct: a report on developments in the availability, marketing and regulation of genetics tests supplied directly to the public, Human Genetics Commission, London. (2007) (URL: http://www.hgc.gov.uk/Client/document.asp?DocId= 139&CAtegoryId=10). Accessed 25 April 2010.

Holtzman, N.A. Promoting safe and effective genetic tests in the United States: work of the task force on genetic testing, Clinical Chemistry 4(5) (1999), 732-738.

International Human Genome Sequencing Consortium (IHGSC) Finishing the euchromatic sequence of the human genome, Nature 431(2004), 931-945. doi:10.1038/nature03001.

Janssens, A.C.J.W., Gwinn, M., Bradley, L.A., Oostra, B.A., van Duijn, C.M., Khoury, M.J. A critical appraisal of the scientific basis of commercial genome profiles used to assess health risks and personalize health interventions, American Journal of Human Genetics 82 (2008), 593-599.

Jordens, C.F.C., Kerridge, I.H., Samuel, G.N. Direct-to-consumer personal genome testing: the problem is not ignorance – it is market failure, American Journal of Bioethics 9(6-7) (2009), 13-15.

Kalf, R., Bakker, R., Janssens, A.C. Predictive ability of direct-to-consumer pharmacogenetic testing: when is lack of evidence really lack of evidence?, Pharmacogenomics 14(4) (2013), 341-344.

Katsanis, S.H., Javitt, G., Hudson, K. A case study of personalised medicine, Science 320 (2008), 53-54.

Kaye, J., Boddington, P., de Vries, J., Hawkins, N., Melham, K. Ethical implications of use of whole genome methods in medical research, European Journal of Human Genetics 18 (2009), 398-403.

Kerr, I., Chandler, J., Caulfield, T.A. Emerging health technologies. In: Caulfield, T.A., Downie J.G., Flood, C.M. (eds.). Canadian Health Law and Policy (4th Ed.), LexisNexis, Canada, Markham, Ontario, 2011.

Khoury, M.J., McBride, C.M., Shully, S.D., Ioannidis, J.P.A. et al. The scientific foundation for personal genomics: recommendations from a National Institutes of Health – Centers for Disease Control and Prevention multidisciplinary workshop, Genetics in Medicine 11(8) (2009), 559-567.

Kraft, P., Hunter, D.J. Genetic risk prediction – are we there yet?, New England Journal of Medicine 360(17) (2009), 1701-1703.

Kuehn, B.M. Risks and benefits of direct-to-consumer genetic testing remain unclear, Journal of American Medical Association 300(13) (2008), 1503-1505.

Kutz, G.D. Direct-to-consumer genetic tests: misleading test results are further complicated by deceptive marketing and other questionable practices, United States Government Accountability Office GAO-10-847T (2010) (URL: http://www.gao.gov/assets/130/125079.pdf). Accessed 8 May 2014.

Landecker, H. Culturing Life: How Cells Became Technologies, Harvard University Press, Cambridge MA, 2007.

Lawrence, F. UK warned of another horse-meat scandal as food fraud rises, The Guardian, 10 October 2013.

Lenzer, J., Brownlee, S. Knowing me, knowing you, British Medical Journal, 336(7649) (2008), 858-860.

McBride, C.M., Wade, C.H., Kaphingst, K.A. Consumers' views of direct-to-consumer genetic information, Annual Review of Genomics and Human Genetics 11 (2010), 427-446.

McGowan, M.L., Fishman, J.R. Using lessons learned from BRCA testing and marketing: what lies ahead for whole genome scanning services?, American Journal of Bioethics 8(6) (2008), 18-20.

McGowan, M.L., Fishman, J.R., Lambrix, M.A. Personal genomics and individual identities: motivations and moral imperatives of early users, New Genetics and Society 29(3) (2010), 261-290.

Metzger, R. Boing Boing's Frauenfelder: Made By Hand. (2010) (URL: http://vimeo.com/12534051). Accessed 11 May 2014.

Meyer, M. Bricoler, domestiquer et contourner la science: l'essor de la biologie de garage, La Découverte I Réseaux. 4(3) (2012), 173-174.

News Medical. 23andMe project aims at validating highly scalable pharmacogenomics research. (2010) (URL: http://www.news-medical.net/news/20101217/23andMe-project-aims-at-validating-highly-scalable-pharmacogenomics-research.aspx). Accessed 25 March 2014.

NHGRI. Frequently Asked Questions About Pharmacogenomics. National Human Genome Research Institute (2014) (URL: http://www.genome.gov/27530645). Accessed 25 March 2014.

Ng, P.C., Murray, S.S., Levy, S., Venter, J.C. An agenda for personalised medicine, Nature 461(8) (2009), 724-726.

OpenPCR. Homepage Open PCR 2013. (URL: http://openpcr.org/>). Accessed 11 May 2014

O'Reilly, T. What is Web 2.0? O'Reilly Media Inc. (2009) (URL: http://books.google.co.uk/books?hl=en&lr=&id=NpEk_WFCMdIC&oi=fnd&pg=PT1&dq=web+2.0&ots=OXQGP2jEI_&sig=gS7Tq4yQ7k3xWT0r3P7DGyS51Io). Accessed 8 May 2014.

O'Riordan, K. The Genome Incorporated: Constructing Biodigital Identity. Theory, Technology and Society, Ashgate, London, 2010.

Oyama, S. The Ontogeny of Information: Developmental Systems and Evolution, Duke University Press, Durham NC, 2000.

Patch, C., Sequeiros, J., Cornel, M.C. Genetic horoscopes: is it all in the genes? Points for regulatory control of direct-to-consumer genetic testing, European Journal of Human Genetics 17 (2009), 857-859.

Pearl Biotech. Home page. (2014) (URL: http://pearlbiotech.com/). Accessed 11 May 2014.

Petersen, A. The Politics of Bioethics, Routledge, New York, 2011.
Pinker, S. My genome, my self, New York Times Magazine. (2009) (URL: http://www.nytimes.com/2009/01/11/magazine/11Genome-t.html?pagewanted=all&_r=0). Accessed 8 May 2014.
Prainsack, B. Voting with their mice: personal genome testing and the participatory turn in disease research, Accountability in Research: Policies and Quality Assurance 18(3) (2011), 132-147.
Prainsack, B., Reardon, J., Hindmarsh, R., Gottweis, H., Naue, U., Lunshof, J.E. Misdirected precaution: personal genome tests are blurring the boundary between experts and lay-people, Nature 456 (2008), 34-36.
Prainsack, B., Wolinsky, H. Direct-to-consumer genome testing: opportunities for pharmacogenomics research?, Pharmacogenomics, 11(5) (2010), 651-55.
Rabinow, P. Artificiality and enlightenment: from sociality to biosociality. In: Biagiolo, M. (ed.). The Science Studies Reader, Routledge, New York, 1999.
Reardon, J. The 'persons' and 'genomics' of personal genomics, Personalized Medicine, 8(1) (2011), 95-107.
Richards, M. Reading the runes of my genome: a personal exploration of retail genetics, New Genetics and Society 29(3) (2010), 291-310.
Roosth, S. Crafting Life: a Sensory Ethnography of Fabricated Biologies, Ph.D. Thesis. Massachusetts Institute of Technology, Boston MA. (2010) (URL: https://dspace.mit.edu/handle/1721.1/63236). Accessed 11 May 2014.
Rose, N., Novas, C. Biological citizenship. In: Ong, A., Collier, S. (eds.). Global Assemblages: Technology, Politics, and Ethics as Anthropological Problems, Blackwell, Oxford, 2004.
Schiele, B., Jacobi, D. La vulgarisation scientifique: thèmes de recherche. In: Schiele, B., Jacobi D., (eds.). Vulgariser la Science: le Procès de l'Ignorance, Syssel, Champ Vallon, 1993.
Shiels, M. Valley girls: Linda Avey and Anne Wojcicki, BBC News Channel, 20 November 2008. (URL: http://news.bbc.co.uk/1/hi/technology/7738768.stm). Accessed 8 May 2014.
Silverman, P.H. Commerce and genetic diagnostics, Hastings Center Report 25(3) Special Supplement (1995), 15-18.
Tocchetti, S. DIYbiologists as 'makers' of personal biologies: how to make Magazine and Maker Faires contribute in constituting biology as a personal technology, Journal of Peer Production 2 (2012) (URL: http://peerproduction.net/issues/issue-2/peer-reviewed-papers/diybiologists-as-makers/). Accessed 11 May 2014.
Tutton, R., Prainsack, B. Enterprising or altruistic selves? Making up research subjects in genetic research, Sociology of Health and Illness 33(7) (2011), 1081-1095.
Udesky, L. The ethics of direct-to-consumer genetic testing, The Lancet 376 (2010), 1377-1378.
Van Ommen, G.J.B., Cornel, M.C. Recreational genomics?, European Journal of Human Genetics 16 (2008), 403-404.
Vorhaus, D., MacArthur, D. Consumer genetics needs more transparency not excessive regulation. Xconomy. (2010) (URL: http://www.xconomy.com/

national/2010/06/23/consumer-genetics-needs-more-transparency-not-excessive-regulation/). Accessed 8 May 2014.

Vorhaus, D. Surreptitious genetic testing: a new bill in Texas and the Iowa straw poll, Genomics Law Report. (2011) (URL: http://www.genomicslawreport.com/index.php/2011/08/12/surreptitious-genetic-testing-a-new-bill-in-texas-and-the-iowa-straw-poll/). Accessed 8 May 2014.

Vorhaus, D. The past, present and future of DTC genetic testing regulation, Genomics Law Report. (2010) (URL: http://www.genomicslawreport.com/index.php/2010/08/05/the-past-present-and-future-of-dtc-genetic-testing-regulation/). Accessed 8 May 2014.

Wade, C.H., Wilfond, B.S. Ethical and clinical practice considerations of genetic counselors related to direct-to-consumer marketing of genetic tests, American Journal of Medical Genetics Part C (Seminars in Medical Genetics) 142(C) (2006), 284-292.

Wallace, H. Most gene test sales are misleading, Nature Biotechnology 26 (2008), 1221.

Ward, M. How the web went world-wide, BBC News (2006) (URL: http://news.bbc.co.uk/1/hi/sci/tech/5242252.stm). Accessed 8 May 2014.

Whalen, J. In attics and closets, 'Biohackers' discover their inner Frankenstein, The Wall Street Journal, 12 May, 2009.

Wolfberg, A.J. Genes on the web: direct to consumer marketing of genetic testing, New England Journal of Medicine, 6(355) (2006), 543-545.

Wright, C., Burton, H., Hall, A., Moorthie, S., Pokorska-Bocci, A., Sagoo, G., Sanderson, S., Skinner, R. Next Steps in the Sequence: the Implications of Whole Genome Sequencing for Health in the UK, PHG Foundation, Cambridge, 2011.

YouScript. What Is YouScript? (2014) (URL: http://youscript.com/healthcare-professionals/what-is-youscript/). Accessed 25 March 2014.

5

Genetic Transparency *versus* Genetic Privacy – The Complex Ethics of Genetic Testing in Humans

Kirsten Brukamp, Gabrielle M. Christenhusz, Caroline Fündling

Genetic transparency and genetic privacy may conflict at times, even though they are not mutually exclusive. The following chapter therefore discusses aspects of these concepts for legal and ethical considerations about genetic testing in humans. The topics include the right to know and the right not to know, access to testing, data confidentiality, autonomy and wellbeing.

Contents and Definitions

Novel genetic analysis tools and efficient genome sequencing technologies promise a seemingly complete insight into the genetic constitution of individuals in the near future. However, this potentially perfect insight may remain an illusory point on the horizon, towards which genetics continually strives but never reaches. It is also debatable whether such an insight is desirable, and if so, to whom. Some restrictions on who is privy to look into the genetic make-up of individuals would seem to be in order, with appropriate legal and ethical underpinnings.

The current chapter seeks to answer these and similar questions by focusing on the concepts of "genetic transparency" and "genetic privacy". We understand these concepts to be related to the issue of access to genetic information: *genetic transparency* refers to access to genetic information by the individual or other groups (e.g. relatives, friends, physicians, employers, insurers, the government); *genetic privacy* refers to withholding this access for oneself or from the parties mentioned above.

Genetic transparency and genetic privacy do not necessarily conflict; one can know one's genetic profile and choose to keep it private. More interesting is the interaction between genetic transparency and genetic privacy in relationship to others. For instance, an individual can be unwillingly genetically transparent to others (e.g. one's family, who share one's genetic material to some extent), and thus not have full control over one's genetic privacy. Sometimes, one can give up one's genetic privacy to enable others to become genetically transparent for the sake of their health.

This chapter explores possible tensions between genetic transparency and genetic privacy from normative, i.e. ethical and legal, perspectives. The first section discusses the two concepts of the "right to know" and the "right not to know", two rights related to the concepts of genetic transparency and genetic privacy. Both rights may be understood as strengthening the individual's options in the decision-making process for or against certain formats of genetic testing. This part also reports on the contemporary situation in the German jurisdiction, which serves as an example of a Western legal system with traditions that support both the right to know and the right not to know. The second section focuses on the implications of predictive genetic testing for health promotion and disease prevention. This type of predictive genetic testing is not devoid of ethical issues. The general topics of genetic transparency and genetic privacy reverberate here in more specific examples of ethical and social concern, e.g. access to genetic testing, and data protection. These more specific issues can facilitate or hinder genetic transparency or genetic privacy, depending on the context. The third section examines the concepts of autonomy and wellbeing with regard to genetic transparency and privacy, focusing particularly on how these traditional concepts are challenged by the use of new genetic technologies. Section three concludes that it is both possible and paramount to strike the right balance between seemingly opposing ethical concepts.

The Right to Know and the Right Not to Know[1]

Introduction

Genetic information is different from other types of medical information: it enables the detection of genetic disorders or an individual's hereditary disposition towards certain diseases, e.g. breast cancer or Alzheimer's disease. It also has a significant impact on family members who share the same genetic endowment.[2] The carrier status of a hereditary disease may influence one's decision to have children. In contrast to other medical diagnoses, genetic testing is mainly predictive. As a result, while still feeling healthy, the individual is faced with the possible onset of hereditary disease in the future.

Being aware of such facts could cause personal harm. This kind of genetic information is often said to have the potential to destroy someone's life plans and to carry serious psychological consequences with it.[3] On the other hand, genetic testing may bring clarity with regard to one's own genetic make-up.[4] The results might free the individual from the burden of uncertainty about genetic disorders, such as Huntington's chorea. Genetic testing may also be used to detect genetic risk factors for multifactorial diseases - people at risk would thus be able to adapt to these conditions.[5] Appropriate prevention will help to relieve the pressure on health services.

As a result, *the right to know* and *the right not to know* have become key terms in the discussion of how to deal with genetic data. *The right to know* is defined as the right to know one's genetic status, and *the right not to know* is defined as the right not to know one's genetic status.[6] With respect to the terms *genetic transparency* and *genetic privacy*, a closer look at the rights to know and not to know shows how difficult it is to weigh up the various interests of the persons and institutions involved in dealing with genetic information. In this respect, *privacy* means the right to be left alone, which includes the

---

[1] This part was contributed by Caroline Fündling.
[2] Wiesemann, 2011, 216.
[3] Taupitz, 1998, 594.
[4] Wiese, 1991, 482.
[5] DFG, 2003, 19ff.
[6] Bund-Länder-Arbeitsgruppe, 1990, 12.

right to keep certain information from disclosure to other people.[7] *Transparency* concerns the issue of access to genetic information, by the affected person or his or her relatives, as well as by private or public institutions.

The following part deals with the development of both rights from ethical principles and their impact on legislation, focusing on the German discussion and the German legal situation. A brief overview is provided on the international relevance of the discussion and the domestic regulations of these rights in Austria and Switzerland. The framework of both rights under German law is described, as well as their connection to the principles of *informed consent*. Finally, selected case studies illustrate how the German jurisdiction has responded to contemporary genetic issues.

Development and International Background of Both Rights

At first, the right not to know played a dominant role in the discussion of a "right to know" and a "right not to know". These terms arose in the early 1980s from the ethical discourse on genetic technology, which included reflection on its impact on human genetics.[8] At the time, many people were concerned about the risks of the new technology. They feared that a (genetically) transparent person, known in German as 'Der gläserne Mensch' (literally translated as 'glass human being'), might become a real prospect in the near future.[9] In the German discussion of life science ethics, the philosopher *Hans Jonas* was the first to mention "a right not to know" in the context of the reproductive cloning of human beings (a fictional scenario). He exposed the problem of the cloned person "knowing too much about himself" as well as others having too much information about him, which, in turn, would lead to a kind of determinism. In this context, a right not to know might help the individual to live his life free from burdensome knowledge, with every day coming as a surprise.[10] As human genetics research progressed, genetic testing became less expensive, thus facilitating the

---

[7] Rothstein, 1994, 133.
[8] Stockter, 2011, 28.
[9] Loretan/Luzatto, 2004, 48.
[10] Jonas, 1985, 189ff.

collection of increasing amounts of information about a person's health. This shifted the debate from a focus on "a right not to know" to "a right to know" as less expensive genetic tests were more widely accessible. The idea of improving a person's health by knowing his or her genetic predispositions, enabling early and targeted prevention, became paramount. This concluded in a claim for "a right to know", including personal access to genetic testing.

According to Beauchamp and Childress's principles of biomedical ethics, the rights to know and not to know can be considered an expression of an individual's autonomy.[11] Autonomy forms one of their four principles of medical ethics: (1) respect for autonomy, (2) non-maleficence, (3) beneficence, and (4) justice.[12] It is important to note that this kind of principlism is just one of the possible ethical approaches, but in medical ethics it is one of the most widely known and accepted models. *Respect for autonomy* is understood as a force opposing paternalistic attitudes and practices in the context of medical treatment, and it requires the patient's aims, wishes and moral concepts to be taken into consideration.[13] On the one hand, *respect* in the context of autonomy understood in a negative sense means taking a decision free from control or manipulative influence. On the other hand, *respect* understood in a positive sense means active enhancement of the patient's ability to make decisions.[14] According to this, health care professionals are obliged to respect the patient's autonomous decision and to inform him or her about the intended medical treatment in order to guarantee a free choice.[15] One expression of the principle of respect for autonomy is the principle of informed consent.[16] The right to know is an expression of the patient's autonomy to obtain access to his or her genetic make-up and – based on this knowledge – to take further decisions about personal health and lifestyle.

In contrast, the right not to know allows the patient to refuse to be informed about his or her genetic constitution. The right not to

---

[11] Andorno, 2004, 436.
[12] Beauchamp/Childress, 2009, 99ff.
[13] Wiesing, 2008, 31.
[14] Ibid.
[15] Ibid.
[16] Ibid.

know has also been regarded as an expression of the patient's autonomy, although waiving the right to be informed initially appears to contradict autonomy.[17] Including the right not to know as an expression of autonomy is convincing since autonomy is understood as the person's ability to make a free choice about disclosure of medical information. Autonomy in this respect does not necessarily require being fully informed - knowledge of the *potential* outcome of a genetic test is sufficient to choose between disclosure and non-disclosure. Thus, the decision not to know is a decision about dealing with (harmful) information, and it has to be respected by the doctor.

The right not to know could - in a wider sense - also be understood as an expression of the principle of non-maleficence, which is an expression of the principle of medical ethics "*primum non nocere*" ("first do not harm"). Accordingly, the doctor must refrain from treatments or actions that may harm the patient.[18] Disclosure of harmful information without consent could be said to infringe this principle.

At the same time in some instances maintaining a right not to know could result in harm, so that non-disclosure of information could sometimes infringe the same principle of non-maleficence.[19] If the genetic test reveals a genetic disorder and serious harm can be avoided by a special treatment or therapy, but the patient decides not to have the test results disclosed, the principles of autonomy and non-maleficence come into conflict. Legislation tries to take into account the implications of conflict between different ethical principles (see below).

The legal discussion of the rights to know and not to know, in the first instance, dealt with legal issues of genetic testing in the context of insurance and employment relationships. On the one hand, insurers and employers are interested in using information obtained from genetic testing to avoid (primarily) financial risks. On the other hand, the individual might be coerced to undergo genetic testing before entering into an employment relationship or during the application for an insurance policy, and in particular a life insurance

---

[17] Andorno, 2004, 435.
[18] Andorno, 2004, 437.
[19] Wilson, 2005.

policy.[20] People were concerned that individuals carrying genetic disorders or hereditary diseases would be discriminated against.

A broad consensus has been reached: (i) The employee's right to privacy takes precedence over the business interests of the employer. In the context of employment relationships, genetic testing is applicable in the framework of an employment medical examination only if the employee is exposed to health hazards from activities at the workplace as a result of a certain genetic disposition.[21] (ii) It was agreed that in the field of non-obligatory insurance (private health, accident or life insurance) insurers should in general not be allowed to require results obtained from predictive genetic testing. Moreover, in 2001, the German Insurance Association committed to not using genetic data by agreeing on a voluntary moratorium that remained effective until 2011. The use of genetic data in insurance relationships is now - with some strict exceptions - prohibited by law. (iii) To prevent discrimination and according to the principle of solidarity, genetic testing would have no impact on compulsory social or health insurance.[22]

The ongoing discussion thus shifts from employment and insurance issues, which is now regulated by the 2009 German Genetic Diagnosis Act (GenDG)[23] in Germany, to the individual's interests and those of the affected family members. Relatives might have an interest in gaining access to test results. "Genetic privacy" plays an important role in this respect. On the one hand, knowing their genetic status may help affected relatives avoid physical harm by taking preventive or therapeutic measures, but on the other hand, relatives might prefer not to be exposed to unwarranted information.[24] This especially applies to late onset and incurable diseases. So one person's wish to safeguard their personal autonomy might clash with the autonomy of other family members: claiming the right to know for one relative might come into conflict with another relative's right not to know.

---

[20] DFG, 2003, 49ff.
[21] DFG, 2003, 50ff.
[22] DFG, 2003, 52ff.
[23] German Genetic Diagnosis Act, 2009.
[24] Laurie, 1999, 129.

A similar issue occurs in the doctor-patient relationship. The concept of confidentiality, aiming at maintaining the patient's privacy, promotes autonomous choices and is therefore an expression of the principle of respect for autonomy.[25] Physicians who treat several family members come into conflict with respecting their patients' autonomy and the duty to act beneficently in the patients' best interest. For example, genetic testing of a mother might discover a genetic disorder or a disease that is easy to cure or prevent, but will have severe consequences if she remains untreated. She refuses to inform her daughter, who might also be affected. Due to the provisions on medical secrecy, the health care professional is not allowed to disclose the patient's genetic information to third parties, including family members. According to the principle of beneficence, one might assume the physician has an obligation to inform the daughter, but respect for autonomy and privacy prohibits the disclosure. So far, a satisfactory solution to the conflicting rights to know and not to know has not been found.

The "right to know" has also played a role in the discussion on personal access to genetic testing in general. It was agreed upon that a general ban on genetic testing could not be justified.[26] Moreover, individuals should be entitled, if they wished, to request their own genetic information. Reliable genetic testing is essential to avoid burdening the patient with false and misleading test results. Thus, genetic testing should be carried out only in certified laboratories that work in conformance with controlled quality standards; this is now regulated by law. It has been argued that genetic testing should only be carried out by health care professionals.[27] In order to help individuals decide whether or not to undergo genetic testing, there was a call for non-directive genetic counselling, which includes discussion of the personal, social and physical consequences of genetic diagnosis.[28]

---

[25] Pantilat, 2008.
[26] Moeller-Herrmann, 2006, 124.
[27] McNally/Cambon-Thomsen, 2004, 54ff.
[28] McNally/Cambon-Thomsen, 2004, 56ff.

International Views on Both Rights

From an international perspective, dealing with genetic information was mentioned by the UNESCO *Universal Declaration on the Human Genome and Human Rights*[29] and the European *Convention on Human Rights and Biomedicine,*[30] both adopted in 1997. Article 5 (c) of the UNESCO Declaration states: "The right of each individual to decide whether or not to be informed of the results of genetic examination and the resulting consequences should be respected." Article 10.2 (of Article 10: private life and right to information) of the European Biomedicine Convention includes some kind of right to know: "Everyone is entitled to know any information collected about his or her health. However, the wishes of individuals not to be so informed shall be observed."

In 2008, the *Additional Protocol to the Convention on Human Rights and Biomedicine, concerning Genetic Testing for Health Purposes*[31] was passed. Article 16.2 put the "right to know" in concrete terms: "Everyone undergoing a genetic test is entitled to know any information collected about his or her health derived from this test. The conclusions drawn from the test shall be accessible to the person concerned in a comprehensible form." The European Biomedicine Convention emphasises the right to know, while disputing the right not to know as a *right* and recognising it merely as a *wish* not to be informed.

The World Health Organization (WHO) also mentioned a right to know and a right not to know in its *Review of Ethical Issues in Medical Genetics*, which was published in 2003.[32] Full disclosure of relevant information concerning an individual's health condition, including genetic status, should facilitate free and informed choice. Informed choice includes new, controversial or ambiguous interpretations of the (genetic) test results. Even if the disclosure may cause anxiety or distress, it is seen as the preferable option in comparison to non-disclosure in order to help a person make an informed decision. The merits of a right not to know are recognized, but people should also

[29] UNESCO, 1997.
[30] Council of Europe, 1997.
[31] Council of Europe, 2008.
[32] WHO, 2003.

understand the extent of information they choose not to know. Selective disclosure of genetic information is not seen as being in the person's best interest, and it is also rather difficult to put into practice, especially when considering the increasing amount of knowledge about genetic disorders. As a result, the WHO document suggests that people should not be encouraged to claim their right *not* to know.

The (partly) German-speaking countries Austria and Switzerland have all set in place a legal framework for genetic testing. The 1994 Austrian Gene Technology Act (GTG) entered into force in January 1995.[33] The Act explicitly regulates genetic examinations of humans for medical purposes. According to the GTG, any subject person is allowed to refuse the disclosure of the test results at any time, which is an expression of the right not to know. The right to know is implied by providing a duty to inform subjects about the nature and extent of the intended genetic examination, and by mandatory genetic counselling.

In Switzerland, genetic testing of humans is regulated by the 2004 Federal Act on Human Genetic Testing (GUMG).[34] Article 6 of the GUMG states a right not to know: "Every person has the right to refuse to receive information about his or her genetic status." Article 18 of the GUMG regulates the person's right to self-determination with regard to genetic testing. A person is free to decide whether to undergo genetic testing, whether to know the result of the test, and what conclusions he or she wishes to draw from the test results. This provision is therefore an expression of both the right to know and the right not to know, but in contrast to the Austrian GTG, the right not to know is regulated explicitly.

Legal Foundation and Framework in German Law

German law distinguishes between constitutional law and subconstitutional law. In contrast to the Common law system, where case law or precedent prevails, German law is mainly based on statutes. The 1949 German Constitution states the fundamental rights

---

[33] Austrian Gene Technology Act, 1994 (GTG - German abbreviation).

[34] Federal Act on Human Genetic Testing, 2004 (GUMG - German abbreviation).

and values that bind state authority and, notably, includes human dignity as the overarching principle. Any subconstitutional law has to be consistent with fundamental constitutional rights and values. The state, and in particular parliament, is responsible for balancing the personal rights derived from the Constitution against each other and public interests by legislation in the respective field of law, e.g. civil law. Constitutional rights and values are used to interpret legal provisions or the extent of a claim.

Both rights, the one to know and the other not to know, are based on the so-called General Right of Personality, which includes the rights to privacy and self-determination.[35] The German Federal Constitutional Court derived this right from Article 2, paragraph 1 (general freedom of action) in conjunction with Article 1, paragraph 1 (human dignity) of the German Constitution. Constitutional rights are primarily meant to guarantee a minimum of defence against state authority. Citizens should be protected against unjustified interference with their personal rights by government agencies.[36] Furthermore, the state has a duty to protect civil rights against threats by third parties. The state may fulfil that duty through legislation.[37]

"Informed consent" is based on the ethical principle of respect for autonomy[38] and - from a legal point of view - on the patient's (constitutional) right to self-determination. Therefore, "informed" in this context means that, before undergoing medical treatment of any kind, the attending physician is obliged to inform the patient of the nature, extent and risks of the intended examination or therapy, possible findings, diagnoses and alternative treatment options. This applies to all stages of the treatment. The amount of information disclosed by the health care professional depends on the respective patient and the level of urgency of the medical intervention. The more urgent the medical intervention, the less information is required to fulfil the duty of informed consent. Identifying the scope of information that the patient wants and is able to deal with is the

---

[35] Scherrer, 2012, 272ff.
[36] Hömig, 2010, 40.
[37] Hömig, 2010, 45.
[38] Schöne-Seifert, 2007, 32ff.; see above.

responsibility of the physician. In this context, "consent" means that the patient's active agreement to medical intervention is required.

The definition of the right (not) to know as the right (not) to know one's genetic status does not specify the implementation of these rights. First, a person needs to be informed about the scope of information that could be drawn from the genetic test results. Thus, informed consent with regard to genetic testing needs to be based on a well-informed understanding of the (genetic) facts and the possible interpretation of the results. Therefore, the right to know is partly expressed in legal provisions that guarantee informed consent with regard to genetic testing such as the duty to inform and genetic counselling. The same applies to the right not to know. The person concerned needs to be aware of the relevant information he or she chooses not to have disclosed. The right not to know is partly expressed in legal provisions that entitle a person to refuse consent or to revoke consent already given. According to the above-mentioned legal provisions, the affected person is enabled to assert his or her right to know or not to know.

But informed consent is not a necessary precondition for the right to know or the right not to know. Both rights go beyond the principles of informed consent. While informed consent should guarantee a person's right to self-determination in the case of a specific medical treatment, the right to know includes the right to have genetic information disclosed regardless of the situation in which genetic information is collected or the person or institution who holds the genetic data. The same applies to the right not to know. The person may assert a right to know or a right not to know even if he or she is not aware of precisely the genetic information that is subject to these rights. This becomes relevant in the case of new technologies in human genetics, especially with regard to "next-generation sequencing", which allows the human genome to be sequenced as a whole, and which produces a great deal of genetic data that cannot yet be fully understood and interpreted.

The 2009 German Genetic Diagnosis Act (GenDG) came into force in February 2010 after approximately two decades of discussion. The GenDG provides a legal framework for the right to know and the right not to know. In order to ensure a person's claim for knowing his or her genetic make-up, genetic testing should be provided by health care professionals or, in the case of predictive testing, by specialists in human genetics and other health care professionals with additional qualifications. According to the principles of informed consent, the attending physician is obliged to inform the patient of the nature, meaning and consequences of the respective test, including the right not to know. Genetic testing should be conducted only after active agreement to the test. In order to help the patient deal with the test outcome, genetic counselling should be offered or, in the case of predictive testing, must be offered. Genetic counselling includes extensive information about the potential medical, psychological and social implications of consent to or refusal of genetic testing.

If the test results affect family members, the patient is asked to encourage his or her relatives to undergo genetic counselling as well, in cases where the patient carries a genetic mutation with significance for serious avoidable or treatable diseases.

To protect the patient's right not to know, he or she has the ability to revoke the consent with future effect at any time. To ensure that a patient only undergoes genetic testing by his or her own free choice, a general principle of non-discrimination is regulated. A general ban on requesting or receiving results obtained from genetic testing in insurance and employment relationships complements the general principle of non-discrimination.

As mentioned above, the rights to know and not to know are based on the general right of personality. As an absolute right, the general right of personality enjoys direct protection under the German law of tort. Infringements may result in compensation for damage.[39] For one thing, health care professionals providing human genetic diagnoses are at risk of being charged with violating a person's right not to know if they disclose genetic information against the will of the patient. For another, informing the person insufficiently might infringe the right to know. Dealing with genetic data, physicians are forced to walk a fine line.

Legislative authority has tried to balance the individual's claim to safeguard his or her autonomy and privacy against third parties and public or private interest in genetic information through the adoption of the GenDG. It is noteworthy that the legislator regarded the protection of individual privacy as being more important than the

---

[39] Kern, 2003, 61.

interests of others. Informing (potentially) affected relatives lies within the patient's discretion, which seems to be an unsatisfactory solution for the following reasons: if the patient does not want to tell relatives, these relatives in effect lose their right to know (in relation to the test results of the first individual). For another thing, if an individual is recommended to undergo genetic counselling after his or her relative has had a genetic test, he or she must assume there is a possibility of carrying a genetic disorder. This person *de facto* loses the right not to know.

Case Studies from the German Jurisdiction

Genetic issues have been subject to only very few court cases in the German jurisdiction. Genetic issues have been subject to criminal proceedings only in the case of DNA analysis to identify offenders, i.e. "genetic fingerprinting". The German Federal Constitutional Court stated that detection and prosecution of crime and criminal offences are paramount and denied an offender's right to refuse the test because of his or her right of personality. The latter is actually affected, but the public interest in criminal prosecution prevails.[40] Two administrative and civil court decisions illustrate the legal consequences of the right to know and the right not to know. The first decision is implicitly based on the right not to know, and the second explicitly mentions the right not to know. The right to know was only subject to court rulings that dealt with the question of paternity. Courts affirmed a person's right to know his or her own biological descent.[41]

The first case example is a judgment of the Administrative Court of Darmstadt from 2004.[42] The facts of this case were as follows: the plaintiff was a teacher who had recently finished her studies, and was teaching at a middle school. She applied for civil servant status. The mandatory medical assessment, including a physical examination turned up cases of Huntington's disease in her family. As a result, she carried a 50% risk of having inherited this genetic disorder. She refused to undergo a genetic test. The state education office rejected

---

[40] BVerfG NJW 1006, 771 (772).

[41] BVerfGE 79, 256.

[42] VG Darmstadt, NVwZ-RR 2006, 566.

her application for civil servant status. The plaintiff entered an objection that was also rejected. In its ruling on the objection, the state education office stated that the carrier status of a hereditary disease constitutes an obstacle for civil servant status. The potential onset of Huntington's chorea poses too high a risk for permanent disability and the related costs for the state. The plaintiff sued the state and claimed her appointment for civil service.

The court decided in favour of the plaintiff. The state had to accept her application for civil service. Affirming the admissibility of a mandatory physical examination with regard to application for civil service in general, the court denied health reasons for a refusal to grant the application in the present case. The physical examination aims to determine the current medical condition and to forecast the future health condition of the candidate, who has an obligation to co-operate. This obligation is limited by a person's general right of personality (which includes a right to privacy) and the principle of proportionality. A genetic test constitutes a serious interference with the applicant's right of personality. She could not be forced to undergo genetic testing to clarify her carrier status for Huntington's disease. In this regard, a 50% risk may not justify a rejection of her application for civil service due to health reasons. The court did not explicitly refer to the plaintiff's right not to know. But de facto, the reasons for the decision suggest such a right not to know, based on the applicant's right of personality as a limit for public information requirements. Thus, both the respect for autonomy and the wish to keep genetic information private prevail in this respect. According to § 19, 22 GenDG, genetic testing before or after the formation of an employment relationship is now prohibited (which also applies to civil service).

The second case is a judgement of the Regional Appeal Court of Rhineland-Palatinate in Koblenz, Germany, from 2013.[43] The facts of this civil case were as follows: the defendant was the senior health care professional of a psychiatric hospital. He was the attending physician of the plaintiff's divorced husband. The defendant diagnosed Huntington's chorea in the husband. At the request of his patient the defendant informed the plaintiff about the diagnosis. The

---

[43] OLG Koblenz, GesR 10/2013, 612.

diagnosis meant that the two mutual children of the plaintiff and her divorced husband bear a 50% risk of carrying the mutation themselves. Pursuant to section 14 GenDG, genetic testing of minors for incurable or untreatable hereditary diseases, such as Huntington's chorea, is not permitted under German law. Section 14 GenDG is generally applicable if a person is not capable of recognizing the nature, meaning or scope of a genetic examination, and is therefore not able to consent to the genetic examination, either because of age or because of mental restrictions. Due to this difficult situation, the plaintiff suffered from severe psychological problems. She therefore asserted a claim for damages and for pain and suffering against the physician as the defendant. Her claim was refused by the competent Regional Court. The Regional Court justified its decision by a "right" or, at least, a legitimate interest of the defendant to inform the plaintiff about the test results of her divorced husband.

The Regional Appeal Court decided in favour of the plaintiff. It affirmed a bodily injury caused by the disclosure of the divorced husband's carrier status for Huntington's chorea. The court admitted a claim for damages and a compensation for pain and suffering. It denied the physician's "right" to inform his patient's divorced wife. There were neither legal reasons nor legitimate interests for such a "right" to inform. A duty of the defendant to inform the two children (as directly affected by the test results) was denied as well, because of the incurability of the disease and the lack of treatment options. Neither the plaintiff nor her children were in a treatment relationship with the defendant. The mere wish of the patient to have his former wife and his children informed may not justify a right or a duty to disclose the test results.

Because of the inability to have her children tested, the plaintiff experienced the carrier status of her divorced husband as a heavy burden. As a result, the court affirmed the plaintiff's and her children's right not to know in the case of incurable hereditary diseases: "the person has a constitutional right not to know his or her carrier status for a hereditary disease." This right is particularly intended to protect affected relatives against the disclosure of (unwanted) genetic information without being asked. The court also stressed the fact that infringements of the right not to know are irrevocable. In this case, the reason for stating a right not to know was not primarily respect for the plaintiff's or her children's autonomy, but the specific content of the genetic information, namely the possibility of carrying a gene causing a disease that leads to suffering and early death.

The German Federal High Court was concerned with that case because the defendant lodged an appeal.[44] The Court again denied the plaintiff's claim for damages. A "right not to know" was also explicitly recognized. The court stated that the right not to know is the right of the person concerned, so the plaintiff was not allowed to claim her children's right not to know.

Conclusions

The rights to know and not to know should be taken into account in discussions on the future of human genetics. At present, the extent of both rights remains unclear. This becomes evident when dealing with genetic information in relation to affected family members. The conflicting rights need to be weighed against each other. An adequate solution, balancing a person's right to know with the relative's right not to know and *vice versa*, has not yet been found. Little available case law makes it difficult to provide physicians with precise instructions to avoid liability risks in the complex field of human genetics.

Ongoing developments in human genetics involve both opportunities and risks. On the one hand, it is important that people benefit from human genetics research and the improvement of genetic testing through new technologies. Therefore, early detection and diagnosis of certain diseases could help to avoid health hazards and to improve a patient's physical comfort. It is probable that future medical treatment will shift from therapeutic to preventive measures. The right to know will become increasingly important. As a comprehensive claim for information, it enables people to improve their health and to participate in technical and medical progress. In this respect, transparency actually becomes an opportunity and an advantage for the individual.

On the other hand, in order to protect the individual's autonomy and privacy, a decision to waive any genetic testing must be accepted. Although legal provisions limit access to genetic information by public authorities, employers and insurers, genetic data remain

---

[44] BGH, VersR 2014, 891.

sensitive, bearing the potential risk of harm to or discrimination against individuals or even whole families. Refusing the disclosure of genetic information is an expression of fundamental human rights because respect for human dignity prohibits treating any human being as an *object*. In this regard, humans as *subjects* have a vital need to keep specific information private to protect their personality. Thus, the right not to know should remain a right recognized by legislation and jurisdiction. In all cases, the law should protect the most vulnerable.

Predictive Genetic Testing for Health Promotion and Disease Prevention[45]

Overview

Contemporary opportunities in genetic testing include the possibility of examining the genomes of individual human beings for susceptibilities to diseases. Despite the promises of genetic testing for health promotion, a number of ethical concerns mean this type of use cannot currently be recommended. The ethically relevant topics include evidence-based practice, access to genetic testing, direct-to-consumer marketing, data protection and confidentiality, genetic counselling, and risk communication. These problems are discussed below because they need to be addressed in the face of increasing usage of genetic testing by consumers worldwide, in part due to direct-to-consumer advertising.

The overall topics of genetic transparency and genetic privacy reverberate in these specific examples of ethical and social concerns. Personal or societal (such as legal and political) influences on the more tangible issues can facilitate or hinder genetic transparency or genetic privacy, depending on the circumstances. For example, policies on data protection and confidentiality may help the individual to gain better control of his or her data and thereby configure genetic privacy or transparency in a more purposeful manner.

---

[45] This part was contributed by Kirsten Brukamp.

Types of Genetic Testing

In this section, the term "genetic testing" refers to the current methods of analysing deoxyribonucleic acid (DNA) material that are performed to make results available to individual clients from whom the material is taken, either directly to consumers or to patients via the health care system. DNA testing is often restricted to finding mutations and variants in genes whose functions are already known. The term "genetics" usually refers to the study of single genes and the associated phenotypes, whereas the discipline of "genomics" concerns analysis of the whole genome, e.g. through genome sequencing and genome-wide screening. The methods used in these types of test differ, but both types of analysis are becoming increasingly widespread because of the decreasing costs of technological equipment.[46] Below, the expression "genetic testing" refers to both genetic and genomic types of testing.

Classical genetic testing in the medical setting in order to establish a diagnosis is normally performed for only a few genes at a time. It confirms a suspected disease process and may therefore be termed diagnostic genetic testing. Here, one of the main ethical issues is the informed consent that the patients need to give. In comparison, this medical setting is much simpler than the application of genetic testing for health promotion and disease prevention, which may be called predictive or presymptomatic. In this case, many more genes or even the whole genome may be checked, even though it is in principle possible to perform targeted predictive testing for only a few genes or diseases at a time. The levels of prevention aimed for by genetic testing in its current format are primary and secondary prevention, i.e. testing conducted to avoid the occurrence of a disease altogether or to diagnose and treat a disease at an early stage so that it does not cause harm. Currently, the role of genetic testing in tertiary prevention, where the focus is on preventing a recurrence of previously treated diseases in patients, remains less clear.

Genetic testing will become increasingly prevalent and popular in the future for several reasons: the costs of whole-genome sequencing

---

[46] Wetterstrand, 2013.

are dropping,[47] genetic knowledge is improving, and public awareness of health and prevention is on the rise. Genetic testing for preventive reasons is associated with the hope that detrimental conditions can be diagnosed before they manifest, disease dispositions could be identified early on so that interventions will be possible, and the administration of medication can be improved through pharmacogenomics.

Genetic testing satisfies the individual's "right to know" about his or her health status. It helps people to live a preventive lifestyle and allows them to be proactive about their health. Nevertheless, genetic testing for prevention carries with it a number of ethical problems, several of which are highly relevant to social debate and political decision-making.

Evidence-Based Practice

Evidence-based practice is an approach to action that emphasizes the importance of empirical findings to justify interventions.[48] It is divided into several branches, including those related to medicine, nursing, and various therapeutic approaches. From the perspective of evidence-based medicine, empirical studies are needed to demonstrate that genetic testing results in decreased mortality and morbidity. Depending on the disease in question, this will be either easy or difficult to show. For example, the complications of one disease may occur in childhood and of another in older age. If testing were commonly performed in childhood, empirical studies would be expected to demonstrate an effect on complications within a few years in the first case, but only after many decades in the second. Occasionally, the goal of optimal health care may be reached by personalized medicine for the individual patient, rather than by the application of general evidence-based guidelines.[49] Consequently, the two concepts sometimes cooperate in synergy and sometimes conflict with each other.

---

[47] Ibid.
[48] Sackett et al., 1996.
[49] Goldberger/Buxton, 2013.

Technological advances occasionally create a hype phenomenon, but it is preferable that they are robustly supported by evidence.[50] Before predictive genetic testing can be justifiably recommended for disease prediction and prevention from a medical standpoint, it needs to be proven effective by the measures of evidence-based medicine. A lack of such evidence has been noted for genomic medicine in general,[51] and in particular for early diagnostic or presymptomatic genetic testing in young people who are strongly predisposed to developing diseases at a later age.[52] Genetic testing has been criticized because of its poor performance in terms of both clinical utility and clinical validity.[53]

Questions therefore arise about whether genetic testing is effective, either in general or for a specific test, and about the purpose for which it is performed, e.g. to serve health or for entertainment as part of popular science. The public interest in health improvement mandates the initiation of studies that prove genetic testing to be helpful across indications, i.e. for different medical diseases, disorders or conditions. This requirement is in line with demands for quality standards in evidence-based medicine. So far, the tests assessed have not been shown to have a relevant impact on public health.[54]

Access to Genetic Testing

Should genetic testing prove successful in promoting health (compare the section above on evidence-based practice), insurance coverage for it is in the public interest. Respecting both patients' and consumers' "right to know", insurance companies may begin to offer genetic testing as part of their services. In doing so, they will promote widespread access to it in some countries. The following ethical issues accompany this development:

Genetic testing might be made available through unequal and therefore unjust routes of access. Some insurance companies may

---

[50] Caulfield et al., 2013.

[51] Manolio et al., 2013.

[52] Duncan/Delatycki, 2006.

[53] Bunnik et al., 2011.

[54] Palomaki et al., 2010; Bellcross et al., 2012.

offer testing for a small variety of select genes, whereas others might sponsor genome-wide screening. Moreover, financial support for the associated genetic counselling (compare the section below on genetic counselling) may differ between insurance companies. If individuals with poor insurance coverage develop an interest in genetic testing, they may need to pay for the tests themselves. Self-paid testing, in turn, may not be affordable for everyone, despite the overall drop in costs of genetic testing.

Differences may occur between social groups in society in their use of genetic testing, thereby invoking questions of justice as an ethical concept. Groups with a higher level of education and a higher income tend to be more aware of genetic testing.[55] Such groups with higher socioeconomic status may benefit more from testing that they initiate themselves as proactive consumers. There are three reasons why this might happen: groups with higher socioeconomic status may be able to afford genetic testing more easily and therefore undergo it more frequently. These groups may also possess greater knowledge about the goals they wish the tests to achieve and how to interpret the results. In addition, they may be better able to invest more financial and time resources into subsequent measures, such as counselling, prevention and treatment. Therefore, inequalities in the opportunities to use genetic testing create ethical challenges to do with considerations of justice.

An easily available offer of access to genetic testing could lead to peer pressure to undergo testing. The mere availability and financial affordability of a test may raise interest and a degree of coercion that would otherwise be absent or negligible. This phenomenon constitutes a threat to exercising one's right not to know.

Moreover, access to genetic testing can be limited due to intellectual property rules.[56] Some health and technology companies hold patents on tests for specific genes. They may be pharmaceutical companies or other companies that perform genetic research and offer products arising from it. They therefore become the sole provider of commercial tests and can set prices independently.[57]

[55] Kolor et al., 2012.
[56] Ormond/Cho, 2014.
[57] Cook-Deegan et al., 2010.

Companies that partially restrict access to genetic testing because of intellectual property rights thereby challenge the individuals' right to know about their health status.

Direct-to-Consumer Marketing

An ethical problem that arises in some jurisdictions is the prevalence of direct-to-consumer marketing. In countries such as the United States of America (USA) and New Zealand, companies specializing in health care advertise their medical products and health-related services directly to consumers. Negative effects have already been noted and reported in the medical literature, for example in Canada. Even though direct-to-consumer advertising is not officially allowed in Canada, it has an effect due to cross-border marketing from the USA. An increase in broadcast advertising was found to result in higher spending on medication in Canada.[58] This effect is likely to be mediated by patients' requests for prescriptions from their physicians.[59]

Several health and technology companies have offered direct-to-consumer genetic testing in the recent past. Examples of former or current health and technology companies that offer direct-to-consumer genetic testing include 23andMe, deCODE Genetics, Navigenics, Gendia, Myriad, Counsyl, GHC, Sciona and Knome.[60] However, some companies have since withdrawn from the market altogether.

A specific example may illustrate the current market difficulties. 23andMe is a health and technology company that has specialized in direct-to-consumer genetic testing since 2006. It had more than 475,000 clients in 2013.[61] A saliva sample is analysed for single-nucleotide polymorphisms (SNP) associated with more than 250 conditions.[62] The company limited its services after a battle with the Food and Drug Administration (FDA), the authority that approves new pharmaceutical products and medical devices in the USA. While

---

[58] Mintzes et al., 2009.
[59] Mintzes et al., 2003.
[60] Howard/Borry, 2013.
[61] Wojcicki, 2013.
[62] Annas/Elias, 2014.

the company had first offered both health reports and ancestry reports, it switched to providing ancestry results only, i.e. reports about one's genetic heritage and ethnic background. This information was prominently disseminated via its webpage: "23andMe provides ancestry-related genetic reports and uninterpreted raw genetic data. We no longer offer our health-related genetic reports."[63] Nevertheless, since then, 23andMe has expanded its services to other markets, including Canada and the United Kingdom.[64]

The FDA in the USA has expressed doubt about the scientific claims of health and technology companies that offer genetic testing for some time, and has asked for more robust evidence.[65] It has been in negotiations with 23andMe since 2009.[66] In particular, it asked for both the diagnostic and the prognostic value of testing to be proven clinically.[67] Overall, criticism is directed at the quality of the reports that the clients received, including results with interpretations of health status and disease risk.

Direct-to-consumer advertising often uses persuasive language as part of an "aggressive marketing" strategy,[68] and even its supporters are currently critical of this strategy:

> 23andMe had previously framed DTC genetic testing as consumer empowerment – giving people direct access to their genetic information without requiring them to go through a physician or genetic counsellor. [...] We think the day will come when this framing is appropriate, but not until the diagnostic and prognostic capability of genomic information has been clinically validated.[69]

The term "health promotion" provides another example of the use of advertising language. The World Health Organization has adopted and supported this expression:

---

[63] 23andMe, 2014.
[64] Mullard, 2015.
[65] Federal Drug Administration, 2011.
[66] Annas/Elias, 2014.
[67] Ibid.
[68] Ibid.
[69] Ibid.

> Health promotion is the process of enabling people to increase control over, and to improve, their health. [...] Health is a positive concept emphasizing social and personal resources, as well as physical capacities. Therefore, health promotion is not just the responsibility of the health sector, but goes beyond healthy lifestyles to wellbeing.[70]

However, the use of this term is not legally protected, and so anyone can refer to it with different interpretations. Health and technology companies relying on direct-to-consumer marketing are likely to extend or contract its meaning according to their business goals.

Direct-to-consumer genetic tests often lack quality standards and regulation.[71] "Although the FDA has long considered health-related genetic tests to be within its jurisdiction, it has not regulated many of them."[72] The detrimental effects on physical and psychological health that would follow from wrong results, and also the financial liabilities that would result from them, remain unclear.

Data Protection and Confidentiality

Genetic testing carries with it all the ethical questions of data protection that have already been described for clinical information and for biobanking. Consumers are concerned about threats to the privacy of information on genetic risk.[73] They also worry about the possibility of genetic discrimination.[74] The risk of subject re-identification persists even for aggregate data.[75] "The ethical issues [include] the confidentiality of data, which, since the genome is personal, can never be anonymised irreversibly and be perfectly safe."[76] The possibilities for future processing of genetic data needs to be a topic in the information and consent process for genetic testing.

---

[70] World Health Organization, 1986.
[71] McGuire et al., 2010.
[72] Zettler et al., 2014, 493.
[73] Goldsmith et al., 2012.
[74] Caulfield et al., 2013.
[75] Dorfman, 2013.
[76] Rehmann-Sutter, 2015.

Insurance companies that pay for genetic testing may be interested in utilizing the results for their benefit, e.g. for decisions on whom to provide with insurance coverage. In some countries, insurance companies are not legally allowed to access client data that originate from health care providers and contain specific medical information, e.g. about diagnoses and therapies. Similar data confidentiality laws still need to be established worldwide. In this setting, it is intriguing to speculate whether expanded personal property rights over one's own genome would help to solve the problems. Such expanded rights have already been proposed in some jurisdictions.[77]

Given the inheritability of DNA and genetic traits, genetic information is always shared. Consequently, "genetic transparency" can conflict with, and is therefore occasionally limited by, the wishes of others for "genetic privacy". Here, "genetic transparency" is understood as knowledge of an individual's genetic composition that is made available to oneself or to others, whereas "genetic privacy" refers to the ignorance about this genetic constitution. The latter value is in danger when patients and consumers openly communicate their genetic information and the associated risks. Their immediate genetic relatives carry some of the same genetic traits. The relatives' decision-making may therefore be compromised, as well as their "right not to know".

Genetic Counselling

The general public may not possess sufficient understanding of the advantages and disadvantages of commercial genetic testing.[78] Patients and consumers should not simply be left with paper or online information on their genetic testing. Both patients and consumers may be confused about the test results, and they may find it particularly hard to draw practical conclusions from reports of numbers or elevated risks. For diagnostic testing, the possibility of a paradoxical effect has been observed where "good" test results have negative consequences because they challenge fundamental plans in

---

[77] cf. The Commonwealth of Massachusetts, 2011-2012.

[78] Dorfman, 2013.

life.[79] For consumers, disappointment ensues when the results are not as exciting as expected. Sometimes, genetic testing is viewed as scientific entertainment, rather than as a helpful tool for health maintenance.

Even for highly trained genetic experts and scholars, personal results may prove ambivalent:

> While this knowledge [about one's own diagnosis] was very helpful overall, it was sobering and ambivalent at the same time to understand my own disease and its implications. The unclarity of my diagnosis somehow had made my life easier, because there was nothing to compare with; the prognosis had been absolutely open. This was exchanged for concerns about possible complications and hopes about potential therapies. However, after all, for me it is better to know rather than to guess about what to expect from my weakened muscles.[80]

Primary care physicians are not qualified to provide highly specialized genetic counselling, and they often do not take enough time for extensive conversations on genetic data. If insurance coverage exists for genetic testing in some countries, it also needs to include expert advice that is capable of communicating and discussing genetic information with the insured persons. This expert advice requires professionals, including clinical geneticists who have been trained as physicians and specialized genetic counsellors who have studied genetic counselling as a topic, but not medicine itself. A quantitative survey revealed that the vast majority of clinical geneticists in Europe are willing to offer genetic counselling to patients after their direct-to-consumer tests, even though these geneticists do not generally support the direct-to-consumer model.[81]

Overall, the resources for such counselling are currently scarce even in countries with well-developed health care systems. For example, the profession of genetic counselling does not exist in Germany, and the German Society of Human Genetics names only a

---

[79] Van Riper, 2005.
[80] Erdmann/Schunkert, 2013, 504.
[81] Howard/Borry, 2013.

few physicians available for counselling purposes in each department for human genetics at university hospitals in Germany.[82] Genetic counselling on test results needs to be standardized across test and service providers in terms of extent, content, availability, cost and consent.

Until recently, systematic empirical data on clients' responses to genetic testing have been scarce, with equally little information on the actions that the clients then take. Such studies are becoming increasingly available over time. One notable counter-example to the relative scarcity of empirical data is the Scripps Genomic Health Initiative,[83] while another study revealed that health behaviours are related to the subjective interpretations and weightings that consumers develop for seemingly objective health data.[84]

The Scripps Genomic Health Initiative,[85] a longitudinal study with more than 2,000 follow-up participants, aimed to examine consumers' responses to direct-to-consumer genetic testing for common diseases. The target domains examined included behaviour, psychology, and health screening. The follow-up health assessment lasted 12 months. In terms of outcomes,[86] about a tenth of the study population had sought advice from a genetic counsellor, and about a quarter had shared the results with their primary physician. No changes from pre- to post-test conditions were discernible in anxiety, test-related distress, diet or exercise. These results indicate a high level of interest in professional counselling, but a low rate of subsequent lifestyle modifications.

A potential hypothesis developed by this study predicts that efforts to provide more extensive, and affordable, genetic counselling might result in a higher rate of lifestyle modifications. If these interventions improved morbidity and mortality, they would generally be judged to be desirable. In consequence, more intense genetic counselling could become an ethically justifiable goal in society.

---

[82] Deutsche Gesellschaft für Humangenetik, 2015.
[83] Bloss et al., 2011.
[84] Kaufman et al., 2012.
[85] Bloss et al., 2011.
[86] Bloss et al., 2011.

Risk Communication

Predictive genetic testing in particular requires education about the statistical concepts of "likelihood" and "risk" for clients of direct-to-consumer testing. Such knowledge is also required for patient education about diagnostic tests in the medical field. Genetic information is not always unequivocal. Risk assessments are part of the scientific approach for medical experts. By contrast, the general public have difficulty in interpreting the science of statistical likelihoods and risks.[87]

> This is a fundamental problem of risk communication because research on risk perception has shown that the understanding of the term 'risk' varies substantially between lay people and scientific experts. While the scientific risk concept is evidence-based and focused, the public addresses uncertainty and a wider range of potential problems.[88]

A qualitative study with interviews and focus groups of professionals and lay people revealed a "need for professional assistance in explaining genetic concepts, including disease causation/risk", which had previously been found in other studies.[89]

For experts, intellectual insights into statistical likelihoods and risks can exert the power to mediate lifestyle changes. Moreover, for experts and lay people alike, subjective interpretations of risk can also be related to health behaviours.[90] In order to respond better to public concerns about disease processes and health risks, it may be helpful to distinguish between "standard epidemiology" and "lay epidemiology", i.e. the scientific epidemiology of experts versus non-scientific observations and assumptions, as well as "between the measurement of risk, which is empirical, and its weighting, which is based on values"[91]. For example, a risk can objectively be low or high, while the acceptance of this risk may paradoxically be low or high for cultural or religious reasons.

---

[87] Hampel, 2006.
[88] Hampel, 2006, 5.
[89] Driessnack et al., 2013, 442.
[90] Kaufman et al., 2012.
[91] Allmark/Tod, 2006.

For genetic counselling, it is a particular challenge to find appropriate ways to communicate statistical likelihoods to patients and consumers so that they can arrive at their own decisions. Several steps in the communication process may be problematic. First, the content of the potential genetic variation needs to be clarified: what is known about the genes in question? Second, it is important to make clear the likelihood with which genetic variants occur so that clients can decide for themselves whether they want to be tested for it (in the case of gene analysis) or receive information about their personal results (for genome-wide screening). Third, the health risks associated with the gene variants need to be described appropriately.

What are the options in risk communication? The general public would probably be overwhelmed if given explanations of absolute risks, relative risks, likelihood ratios, odds ratios, and related concepts from epidemiology and public health, given what we already know about "lay epidemiology".[92] A simpler kind of feedback to clients or patients may either report personal risk in comparison to the general population, or it may reduce the information to designate personal risk as normal, high or low, based on a threshold model that ignores minor changes in risk. While the latter approach is better suited to reaching a larger population, it loses medical information. Simultaneous offers of alternative representations might be least paternalistic because the clients can then choose between them. It is questionable whether sophisticated analyses hold any advantages for individuals, aside from scientific pursuits, when the results cannot be passed on to the potential beneficiaries in order to empower them in their own decision-making processes.

Conclusions

Consumer-initiated genetic testing, as opposed to that recommended by physicians to their patients, is likely to rise in public perception. Advertising efforts through modern media reach not only the target groups in countries where direct-to-consumer marketing is formally legal, but also potentially a worldwide audience across countries and jurisdictions. Accordingly, the ethical, legal and social issues

---

[92] Hampel, 2006; Dorfman, 2013.

associated with genetic testing need to be addressed. To solve them, both empirical data and policy development[93] are required in the short and medium term.

In summary, both consumers and patients exercise their "right to know" and their "right not to know" by initiating client-driven genetic testing, either directly or via their physicians, or by ignoring or refusing it. Ethical reflection reveals that genetic testing presents serious problems. The specific issues that have been discussed here include evidence-based practice, access to genetic testing, direct-to-consumer marketing, data protection and confidentiality, genetic counselling, and risk communication.

All of these topics are related to the overarching themes of "genetic transparency" and "genetic privacy". The problems influence individuals' decision-making processes for knowledge versus ignorance about oneself or others. This becomes particularly apparent with reference to potential interventions, which are agreed upon in society. For example, a stronger focus on evidence-based medicine may result in high-quality empirical evidence that predictive genetic testing for a condition is either beneficial or not in term of morbidity and mortality. Communicating such data to citizens will enable them to evaluate the information for themselves and to make informed decisions on how to respond.

Respect for Autonomy and Wellbeing in the Light of Genetic Transparency and Genetic Privacy[94]

Two ethical concepts that often appear to be contradictory in reflections on the new genomic era are wellbeing and respect for autonomy. The first section of this chapter introduced the concept of autonomy as the basis of respect for a right (not) to know, before elaborating the legal basis and consequences of the right to know and the right not to know. This section explores further some of the complexities of respecting autonomy in the context of new genetic technologies, as well as challenges to safeguarding wellbeing. The

[93] McGuire et al., 2010.

[94] This part was contributed by Gabrielle M. Christenhusz.

section ends by considering these concepts in the light of genetic transparency and genetic privacy, with a reflection on the possibility of striking the right equilibrium. Examples are used where appropriate.

Respect for Autonomy

Respect for autonomy was defined earlier in this chapter, in both a negative and a positive sense.[95] As stated there, respect for autonomy in its negative sense implies being able to make decisions free from control or manipulative influence. In terms of genetic testing using new genetic technologies, a negative understanding of autonomy refers in the first place to being able to make a free decision to undergo a particular genetic test or not. This naturally assumes the possibility of a test. The absence of a particular test - for example, a test that is not freely available in a particular jurisdiction because it is too costly or has yet to be validated, or particular predictive genetic tests in the case of children - rules out the possibility of choosing testing or not. In the second place, respect for autonomy understood in this negative sense refers to being able to make decisions freely based on the results of a particular genetic test.

The exercise of autonomy will be restricted by the range of decisions considered possible; the range of decisions is often only understood in a clinical sense. For instance, if the result of a genetic test indicated an increased risk of late-onset Alzheimer's disease, there is very little in a clinical sense that one could do with this information, certainly in countries where advanced directives for euthanasia are prohibited. The disclosure or nondisclosure of such a result would seem to be neutral with respect to autonomy, if we remain in clinical terms. However, if we look more widely than the strictly clinical, to encompass possibilities for practical and inner preparation, the disclosure of such a result would seem to respect autonomy.

This leads to a consideration of respect for autonomy in a positive sense: the active enhancement of a person's ability to make decisions. This can again be divided into two types of decisions: the decision to

---

[95] cf. the section "The Right to Know and the Right Not to Know".

undergo a particular genetic test or not, and the decisions that then stem from knowledge of the test results. What makes the new form of genetic tests such as whole-genome sequencing (WGS) so challenging is their untargeted nature. Previous genetic tests were targeted to single conditions, and hence mutations in one or a limited number of genes. The informed consent process and accompanying genetic counselling simply had to focus on this single condition, and the implications for the individual and their family of knowing that they had this condition. There is talk of a "paradigm shift" when comparing genetic testing in the new genomic era with previous modes of genetic testing, in terms of the amount of genetic data produced and the bioinformatics and other support needed to use and interpret this data to the fullest extent.[96] The result is that the clinical geneticists, laboratory personnel, and patients may not completely know what they are saying "yes" to in ordering a particular genetic test such as WGS.

Alternative strategies for coping with this mass of potential data range from the development of filters in an attempt to limit possible results to only what is clinically relevant,[97] to the deliberate screening of all known pathogenic, clinically actionable genes alongside the primary clinical question in an attempt to use the beneficial potential of new genetic tests to the maximum.[98] At the very least, what needs to be emphasised to patients is the impossibility of achieving complete "genetic transparency": impossible at the present time because of limitations in knowledge, and for all time because of the inevitable unpredictability of multiple gene-environment interactions. Furthermore, a probably conservative estimate states that a discussion of all possible examples of genetic test results and disclosure and follow-up options with patients could take hours.[99] The question is whether such extensive counselling is the best way to secure either respect for autonomy or the wellbeing of patients.

The precise relationship between being informed and respect for autonomy is unclear. There are those who argue that the only way to

---

[96] Mardis, 2008.

[97] Christenhusz/Devriendt et al., 2012.

[98] Green/Berg et al., 2013.

[99] Mayer/Dimmock et al., 2011.

respect patient autonomy fully is by providing patients with the maximum amount of choice in pre-test counselling and informed consent discussions regarding the various types of results that they would like returned to them.[100] In such cases, respect for autonomy is equated with being given the maximum amount of information. In contrast, approaches in which the patient is informed that only a limited number of results will be returned to them, results deemed relevant by the clinical geneticist, in a "take it or leave it" kind of informed consent procedure, are judged to be paternalistic according to the liberal understanding of respect for autonomy. However, one question is whether being more informed actually facilitates the making of autonomous decisions. As already stated,[101] it has been generally accepted that the doctrine of informed consent is based on respect for autonomy.[102] Nonetheless, someone's autonomy can be respected and they can decide and act autonomously without being fully informed.[103] Indeed, because it is currently impossible to predict the phenotypic outcome of many gene variants, and it may remain impossible due to the above-mentioned unpredictability of gene-environment interactions, it is questionable whether informing patients of *all* possible results will help them in making an autonomous choice. Second, in answer to the critique of paternalism, it is good to note that while the doctrine of informed consent was originally developed in the context of invasive procedures, it is more closely related to the provision of information in the context of genetic testing.[104] In the context of invasive procedures, medical paternalism can be a statement on the professional-subject relationship, an ethical statement regarding respect or disrespect for autonomy or wellbeing. In more information-related contexts, such as genetic testing, medical paternalism is more a statement about the quality of the information than the quality of the relationship; it is a statement about who is better able to judge the quality of the

---

[100] Siegal/Bonnie et al., 2012.
[101] cf. the section "The Right to Know and the Right Not to Know".
[102] Beauchamp/Childress, 2009.
[103] Taylor, 2004; Walker, 2013.
[104] Veach/Bartels et al., 2001.

information. The significance of much genetic information is uncertain given the current state of knowledge.

A better way to respect the autonomy of patients than providing them with all possible information is to strive to do everything possible to facilitate their ability to choose freely between those options that are welfare-promoting.[105] This is often called "libertarian paternalism", a form of non-coercive paternalism that involves presenting a range of options in such a way that people are more likely to choose what will most enhance their welfare.[106] Libertarian paternalism is becoming increasingly popular in genetics, corresponding with the increasing potential overload of genetic information. There is after all a limit to how much information people can cope with while still being able to make autonomous decisions.[107] This should be taken into account both before a genetic test is carried out and when genetic test results are returned. Unrestricted choice before the test is carried out can be overwhelming,[108] while disclosing too many results can drown out the seriousness of the truly important findings.[109] In both instances, information overload can lead to a kind of paralysis, hindering people from being able to choose anything at all.

Respect for Wellbeing

The precise relationship between being informed and wellbeing is similarly unclear. The most obvious example of when being informed will increase wellbeing is being informed of a genetic condition that is certain to manifest and which can be avoided completely or ameliorated by taking definite steps. However, there are very few genetic conditions that fit this profile. It is mostly a question of increased or decreased risk, with the concomitant danger of over-treatment (or under-treatment).

For instance, the discovery of a mutation in a woman's *BRCA1* gene means an average risk by age 70 of 65% for breast cancer and

---

[105] Sunstein/Thaler, 2003.
[106] Thaler/Sunstein, 2003.
[107] Bredenoord/Onland-Moret et al., 2011.
[108] Bowdin/Ray et al., 2014.
[109] Christensen/Green, 2013.

39% for ovarian cancer.[110] If the woman opts for prophylactic mastectomy and hysterectomy, she might fall within the 35% and 61% of women with a BRCA1 mutation who are destined never to develop breast or ovarian cancer, respectively. Her preventive actions would then result in her needlessly having to undergo the psychological and biological challenges associated with these procedures; needless because she was never destined to develop breast or ovarian cancer. However, this remains uncertain. The same woman with a non-causal BRCA1 mutation might have another as yet uncharacterised mutation related to a higher risk of breast and ovarian cancers; or she might develop breast or ovarian cancer due to malignant environmental factors, completely independent of her genetic predisposition. Because it is difficult to characterise the predictability of the information being disclosed, it is difficult to know what will most enhance wellbeing.

An additional difficulty arises due to the fact that the concept of wellbeing is open to interpretation. For example, it might be assumed that being informed that one is not a carrier for Huntington's disease, unlike the rest of the family, would increase wellbeing by relieving unnecessary anxiety. However, some people then report feeling "survivor guilt",[111] while others may have to adjust after having planned their future assuming that they would develop Huntington's disease.[112] It is advisable to identify, in as detailed a manner as possible, all the elements of the information being conveyed that will impact wellbeing, before attempting to conclude whether wellbeing will be enhanced or diminished.

The biopsychosocial model of genomic conditions developed by Rolland and Williams is one model that tries to elucidate the key characteristics of different types of genetic disorders that will make psychosocial (and biological) demands over time.[113] These characteristics include the probability of developing a condition based on specific genetic mutations, overall clinical severity, and whether effective treatment options exist to affect the condition's onset,

---

110 Antoniou/Pharoah et al., 2005.

111 Huggins/Bloch et al., 1992.

112 Huniche, 2011.

113 Rolland/Williams, 2005.

progression or severity. Significantly, Rolland and Williams pay special attention to the timing of clinical onset in their model, noting that a condition set to manifest in early adulthood and childbearing years will carry with it different psychosocial stresses than a condition of old age, and that the presymptomatic phase will also have its own demands. Only by teasing out all of the different facets of any one piece of genetic information can a reasonable estimate be made of the sum effect on wellbeing.

One intriguing question is how consideration of the wellbeing of others should be included when deciding how to apply genetic information. For instance, some research looks at the familial motivations of patients in applying for genetic testing, motivations that are often caught up in ideas about wellbeing. One example is that women who suspect that they are at risk of breast cancer may apply for genetic testing in order to help their family in two ways: to be able to pass on their genetic result to family members, who can then decide about their own testing, and to undertake recommended preventive measures if needed, again often for the sake of their family.[114] Part of chapter 7 in this book looks at when this free choice to include consideration of the wellbeing of one's family in decisions about genetic testing can become a moral expectation and obligation - when choices about genetic transparency have moral consequences.

A related scenario concerns parents who take the wellbeing of others into account when making choices about their unborn or future children, alongside considerations of the wellbeing of their children.[115] The latter principle is widely known as "procreative beneficence".[116] The former principle, called "procreative altruism", argues that parents should select that child whose existence is predicted to contribute the most to (or detract the least from) the wellbeing of others. Douglas and Devolder admit that it may prove difficult to put procreative altruism into practice for anything but the simplest psychological and physical traits.[117] It is also difficult to see how the principle could be applied except in in-vitro fertilization

---

114 Hallowell, 1999.
115 Douglas/Devolder, 2013.
116 Savulescu/Kohane, 2009.
117 Douglas/Devolder, 2013.

(IVF) contexts where pre-implantation genetic diagnosis (PGD) is available, and these techniques come with their own ethical challenges and critics. However, it remains an interesting question: how far should parents go to ensure that their reproductive actions are not just of benefit to themselves and their children, but also of wider benefit? In general, how should questions of beneficence beyond the individual and the family be taken into account when considering genetic testing? Where do the lines of altruism and justice stop?

Genetic Transparency versus Genetic Privacy

Interestingly, the new genomic era introduces the possibility of conflict between the concepts of genetic transparency and genetic privacy. As defined in the introduction, both concepts are related to the question of access to genetic information: genetic transparency refers to access to genetic information by the individual or others, while genetic privacy refers to withholding this access from certain parties by the individual. This section explores the interaction and possible conflict between genetic transparency towards *others* and genetic privacy. Previously, conducting genetic tests that were each only able to target single conditions could result in a maximum amount of both genetic transparency (confirmation of the presence or absence of a given genetic condition) and genetic privacy if so desired. As genetic research increasingly focuses on more complex genotype-phenotype interactions, it becomes necessary to compare the phenotypes of people with the same genotype to identify which genotypes are harmful and to what extent. The first level that will be affected will be privacy between family members; as Wilfond and Ross state, "intra-familial disclosure may become more necessary because of the need to correlate phenotypes and genotypes both to improve the personal predictions of one's own health status and to improve the predictive value of testing more generally."[118]

One may have to choose between genetic transparency and genetic privacy. This is particularly true in the event of so-called *variants of unknown significance* (VUS), when phenotypic information of

---

[118] Wilfond/Ross, 2009.

other family members with the same genotype can be vital in identifying a genetic variant as pathogenic or benign.[119] This puts a new spin on the autonomy-privacy-wellbeing dilemma. Previously, the question was sometimes whether to respect someone's autonomous choice for privacy over the wellbeing of other family members. A new question arises in the new genomic era: whether to forgo privacy and the autonomous choice of family members not to be tested in favour of the potential wellbeing of all family members, including the original patient.

Complete genetic privacy is difficult to ensure in an age where re-identification of so-called "anonymous samples" is a real possibility.[120] It is also debatable how high the value of genetic privacy should be if the incomplete information resulting from complete privacy can lead to harm, as described above.

Paradoxically, the greater the amount of genomic information to which we have access, the less genetically transparent we may become, relative to the complexity and ambiguity of the ensuing data. We have given a glimpse above of the increasing intricacy of the concepts of respect for autonomy and wellbeing due to the new possibilities offered by new genetic tests. The way forward is to try to find the right balance between all of these concepts in changing circumstances, without letting any single concept become generally dominant.

Literature

23andMe. (URL: https://www.23andme.com). Accessed 1 June 2014.

Allmark, P., Tod, A. How should public health professionals engage with lay epidemiology? Journal of Medical Ethics 32 (2006), 460-463. doi: 10.113 6/jme.2005.014035.

Andorno, R. The right not to know: an autonomy based approach, Journal of Medical Ethics (30) (2004), 435-440.

Annas, G.J., Elias, S. 23andMe and the FDA, New England Journal of Medicine 370 (11) (2014), 985-988.

Antoniou, A.C., Pharoah, P.D. et al. Breast and ovarian cancer risks to carriers of the BRCA1 5382insC and 185delAG and BRCA2 6174delT mutations: a

119 Crawford/Foulds et al., 2013.

120 Malin/Sweeney, 2001; Schmidt/Callier, 2012.

combined analysis of 22 population based studies, Journal of Medical Genetics 42(7) (2005), 602-603.

Austrian Gene Technology Act (GTG) (1994). (URL: http://www.bmg.gv.at/cms/home/attachments/7/8/8/CH1060/CMS1226929588865/510_1994.pdf). Accessed 7 January 2014.

Beauchamp, T.L., Childress, J.F. Childress. Principles of biomedical ethics, Oxford University Press, New York, 2009.

Bellcross, C.A., Page, P.Z., Meaney-Delman, D. Direct-to-consumer personal genome testing and cancer risk prediction, The Cancer Journal 18(4) (2012), 293-302.

Bloss, C.S., Schork, N.J., Topol, E.J. Effect of direct-to-consumer genomewide profiling to assess disease risk, New England Journal of Medicine 364(6) (2011), 524-534.

Bowdin, S., Ray, P. et al. The genome clinic: A multidisciplinary approach to assessing the opportunities and challenges of integrating genomic analysis into clinical care, Human Mutation 35 (2014), 513-519.

Bredenoord, A.L., Onland-Moret, N.C. et al. Feedback of individual genetic results to research participants: in favor of a qualified disclosure policy, Human Mutation 32(8) (2011), 861-867.

Bund-Länder Arbeitsgruppe Genomanalyse. Abschlußbericht, Bundesanzeiger 161a, 29 August 1990, 12.

Bunnik, E.M., Schermer, M.H.N., Janssens, A.C.J.W. Personal genome testing: test characteristics to clarify the discourse on ethical, legal and societal issues, BMC Medical Ethics 12 (2011), 11.

Caulfield, T., Chandrasekharan, S., Joly, Y., Cook-Deegan, R. Harm, hype and evidence: ELSI research and policy guidance, Genome Medicine 5(3) (2013), 21. doi: 10.1186/gm425.

Christenhusz, G.M., Devriendt, K. et al. Why genomics shouldn't get too personal: In favor of filters: Re: Invited Comment by Holly K. Tabor et al. in American Journal of Medical Genetics Part A Volume 155, American Journal of Medical Genetics 158A (2012), 2641-2642.

Christensen, K.D., Green, R.C. How could disclosing incidental information from whole-genome sequencing affect patient behavior? Personalized Medicine (10) (2013), 377-386.

Cook-Deegan, R., DeRienzo, C., Carbone, J., Chandrasekharan, S., Heaney, C., Conover, C. Impact of gene patents and licensing practices on access to genetic testing for inherited susceptibility to cancer: comparing breast and ovarian cancers with colon cancers, Genetic Medicine 12 (4 Suppl.) (2010), 15-38.

Council of Europe. Additional Protocol to the Convention on Human Rights and Biomedicine, concerning Genetic Testing for Health Purposes, Strasbourg (2008). (URL: http://conventions.coe.int/Treaty/EN/Treaties/Html/203.htm). Accessed 7 January 2014.

Council of Europe, Convention on Human Rights and Biomedicine, Oviedo (1997). (URL: http://conventions.coe.int/Treaty/EN/Treaties/Html/164.htm). Accessed 7 January 2014.

Crawford, G., Foulds, N. et al. Genetic medicine and incidental findings: it is more complicated than deciding whether to disclose or not, Genetics in Medicine 15(11) (2013), 896-899.

Deutsche Forschungsgemeinschaft (DFG). Predictive genetic diagnosis, Bonn (2003). (URL: http://www.dfg.de/download/pdf/dfg_im_profil/reden_stellungnahmen/2003/Predictive_genetic_diagnosis.pdf). Accessed 7 January 2014.

Deutsche Gesellschaft für Humangenetik. Genetische Beratung in Klinik und Praxis. (URL: www.gfhev.de/de/beratungsstellen/beratungsstellen.php). Accessed 22 February 2015.

Dorfman, R. Ruslan Dorfman replies, Nature Biotechnology 31 (12) (2013), 1076.

Douglas, T., Devolder, K. Procreative altruism: Beyond individualism in reproductive selection, Journal of Medicine and Philosophy 38 (2013), 400-419.

Driessnack, M., Daack-Hirsch, S., Downing, N., Hanish, A., Shah, L.L., Alasagheirin, M., Simon, C.M., Williams, J.K. The disclosure of incidental genomic findings: an "ethically important moment" in pediatric research and practice, Journal of Community Genetics 4(4) (2013), 435-444. doi: 10.1007/s12687-013-0145-1.

Duncan, R.E., Delatycki, M.B. Predictive genetic testing in young people for adult-onset conditions: Where is the empirical evidence? Clinical Genetics 69 (2006), 8-16. doi: 10.1111/j.1399-0004.2005.00505.x.

Erdmann, J., Schunkert, H. Forty-five years to diagnosis, Neuromuscular Disorders 23(6) (2013), 503-505. doi: 10.1016/j.nmd.2013.03.006.

Federal Act on Human Genetic Testing (GUMG) (2004). (URL: http://www.admin.ch/ch/e/rs/8/810.12.en.pdf). Accessed 7 January 2014.

Federal Drug Administration (FDA). FDA Executive Summary. Molecular and Clinical Genetics Panel. March 8/9, 2011. (URL: www.fda.gov/downloads/AdvisoryCommittees/CommitteesMeetingMaterials/MedicalDevices/MedicalDevicesAdvisoryCommittee/MolecularandClinicalGeneticsPanel/UCM245660.pdf). Accessed 2 March 2013.

German Genetic Diagnosis Act (GenDG) (2009), Federal Law Gazette [BGBl.], Bundesanzeiger I, 2009, 2529-3672.

Goldberger, J.J., Buxton, A.E. Personalized medicine vs guideline-based medicine, JAMA 309(24) (2013), 2559-2560. doi: 10.1001/jama.2013.6629.

Goldsmith, L., Jackson, L., O'Connor, A., Skirton, H. Direct-to-consumer genomic testing: systematic review of the literature on user perspectives, European Journal of Human Genetics 20(8) (2012), 811-816.

Green, R.C., Berg, J.S. et al. ACMG recommendations for reporting of incidental findings in clinical exome and genome sequencing, Genetics in Medicine 15(7) (2013), 565-574.

Hallowell, N. Doing the right thing: Genetic risk and responsibility, Sociology of Health & Illness 21 (1999), 597-621.

Hampel, J. Different concepts of risk – a challenge for risk communication, International Journal of Medical Microbiology 296 (S. 1/Suppl 40) (2006), 5-10.

Hömig, D. (ed.). Grundgesetz, 9th Edition, Nomos, Baden-Baden, 2010.

Jonas, H. Technik, Medizin und Ethik – Zur Praxis des Prinzips der Verantwortung, Suhrkamp Verlag, Frankfurt, 1985.

Huggins, M., Bloch, M. et al. Predictive testing for Huntington disease in Canada: adverse effects and unexpected results in those receiving a decreased risk, American Journal of Medical Genetics 42(4) (1992), 508-515.

Huniche, L. Moral landscapes and everyday life in families with Huntington's disease: Aligning ethnographic description and bioethics, Social Science & Medicine 72(11) (2011), 1810-1816.

Kaufman, D.J., Bollinger, J.M., Dvoskin, R.L., Scott, J.A. Risky business: risk perception and the use of medical services among customers of DTC personal genetic testing, Journal of Genetic Counseling 21(3) (2012), 413-422. doi: 10.1007/s10897-012-9483-0.

Kern, B.-R. Unerlaubte Diagnostik – Das Recht auf Nichtwissen. In: Dierks et al. (eds). Genetische Untersuchungen und Persönlichkeitsrecht, Springer, Berlin, 2003, 55-69.

Kolor, K., Duguette, D., Zlot, A., Foland, J., Anderson, B., Giles, R., Wrathall, J., Khoury, M.J. Public awareness and use of direct-to-consumer personal genomic tests from four state population-based surveys, and implications for clinical and public health practice, Genetics in Medicine 14(10) (2012), 860-867.

Laurie, G. T. In defence of ignorance: genetic information and the right not to know, European Journal of Health Law (6) (1999), 119-132.

Loretan, A., Luzatto, F. (eds). Gesellschaftliche Ängste als theologische Herausforderung, Lit Verlag, Münster, 2004.

Malin, B., Sweeney, L. Re-identification of DNA through an automated linkage process, Proc AMIA Symp (2001), 423-427.

Manolio, T.A., Chisholm, R.L., Ozenberger, B., Roden, D.M., Williams, M.S., Wilson, R., Bick, D., Bottinger, E.P., Brilliant, M.H., Eng, C., Frazer, K.A., Korf, B., Ledbetter, D.H., Lupski, J.R., Marsh, C., Mrazek, D., Murray, M.F., O'Donnell, P.H., Rader, D.J., Relling, M.V., Shuldiner, A.R., Valle, D., Weinshilboum, R., Green, E.D., Ginsburg, G.S. Implementing genomic medicine in the clinic: the future is here, Genetics in Medicine 15(4) (2013), 258-267. doi: 10.1038/gim.2012.157.

Mardis, E.R. Next-generation DNA sequencing methods, Annual Revue Genomics Human Genetics 9 (2008), 387-402.

Mayer, A.N., Dimmock, D.P. et al. A timely arrival for genomic medicine, Genetics in Medicine 13(3) (2001), 195-196.

McGuire, A.L., Evans, B.J., Caulfield, T., Burke, W. Science and regulation. Regulating direct-to-consumer personal genome testing, Science 330(6001) (2010), 181-182.

McNally, E., Cambon-Thomsen, A. The independent expert group. Ethical, legal and social aspects of genetic testing: research, development and clinical applications, European Commission, Brussels, 2004.

Mintzes, B., Barer, M.L., Kravitz, R.L., Bassett, K., Lexchin, J., Kazanjian, A., Evans, R.G., Pan, R., Marion, S.A. How does direct-to-consumer advertising (DTCA) affect prescribing? A survey in primary care environments with and without legal DTCA, Canadian Medical Association Journal 169(5) (2003), 405-412.

Mintzes, B., Morgan, S., Wright, J.M. Twelve years' experience with direct-to-consumer advertising of prescription drugs in Canada: a cautionary tale, PLoS One 4(5) (2009), e5699.

Moeller-Herrmann, M. Die Regelung prädiktiver Gentests, Tectum-Verlag, Marburg, 2006.

Mullard, A. 23andMe sets sights on UK/Canada, signs up Genentech, Nature Biotechnology 33(2) (2015), 119.doi: 10.1038/nbt0215-119a.

Ormond, K.E., Cho, M.K. Translating personalized medicine using new genetic technologies in clinical practice: the ethical issues, Per Med 11(2) (2014), 211-222. doi: 10.2217/pme.13.104.

Palomaki, G.E., Melillo, S., Neveux, L., Douglas, M.P., Dotson, W.D., Janssens, A.C., Balkite, E.A., Bradley, L.A. Use of genomic profiling to assess risk for cardiovascular disease and identify individualized prevention strategies – a targeted evidence-based review, Genetics in Medicine 12(12) (2010), 772-784.

Pantilat, S. Autonomy vs. Beneficence, UCSF School of Medicine, California (2008). (URL: http://missinglink.ucsf.edu/lm/ethics/Content%20Pages/fast_fact_auton_bene.htm). Accessed 7 January 2014.

Rehmann-Sutter, C. The new genomic patient, Swiss Med Weekly 145 (2015), w14089. doi: 10.4414/smw.2015.14089.

Rolland, J.S., Williams, J.K. Toward a biopsychosocial model for 21st-century genetics, Family Process Journal 44(1) (2005), 3-24.

Rothstein, M.A. Genetic Discrimination in Employment Law: Ethics, Policy and Comparative Law. In: International colloquium, Lausanne. Human genetic analysis and the protection of the personality and privacy, Zürich, 1994, 129-141(133).

Sackett, D.L., Rosenberg, W.M., Gray, J.A., Hynes, R.B., Richardson, W.S. Evidence-based medicine: what it is and what it isn't, British Medical Journal 312(7023) (1996), 71-72.

Savulescu, J., Kahane, G. The moral obligation to create children with the best chance of the best life, Bioethics 23 (2009), 247-90.

Scherrer, J. Das Gendiagnostikgesetz, Lit Verlag, Münster, 2012.

Schmidt, H., Callier, S. How anonymous is "anonymous"? Some suggestions towards a coherent universal coding system for genetic samples, Journal of Medical Ethics 38(5) (2012), 304-309.

Schöne-Seifert, B. Grundlagen der Medizinethik, Alfred Kröner Verlag, Stuttgart, 2007.

Siegal, G., Bonnie, R.J. et al. Personalized Disclosure by Information-on-Demand: Attending to Patients' Needs in the Informed Consent Process, J Law Med Ethics 40(2) (2012), 359-367.

Stockter, U. Wissen als Option nicht als Obliegenheit. In: Dutte et al. (eds). Das Gendiagnostikgesetz im Spannungsfeld von Humangenetik und Recht, Göttingen, 2011, 27-51.

Sunstein, C.R., Thaler, R. H. Libertarian paternalism is not an oxymoron, U. Chi. L. Rev. 70 (2003), 1159.

Taupitz, J. Das Recht auf Nichtwissen. In: Hanau et al. (eds). Festschrift für Günther Wiese zum 70. Geburtstag, Neuwied, 1998, 583-602.

Taylor, J.S. Autonomy and Informed Consent: A Much Misunderstood Relationship, The Journal of Value Inquiry 38(3) (2004), 383-391.
Thaler, R.H., Sunstein, C.R. Libertarian Paternalism, The American Economic Review 93(2) (2003), 175-179.
The Commonwealth of Massachusetts. Massachusetts Senate Bill 1080: An act to create a genetic bill of rights, MA S1080, 2011-2012. (URL: legiscan.com/ MA/text/S1080/2011). Accessed 2 March 2013.
UNESCO. Universal Declaration on the Human Genome and Human Rights, Paris (1997). (URL: http://portal.unesco.org/en/ev.phpURL_ID=13177&URL _DO=DO_TOPIC&URL_SECTION=201.html). Accessed 7 January 2014.
Van Riper, M. Genetic testing and the family. Journal of Midwifery and Women's Health 50(3), 227-233, 2005.
Veach, P.M., Bartels, D.M. et al. Ethical and Professional Challenges Posed by Patients with Genetic Concerns: A Report of Focus Group Discussions with Genetic Counselors, Physicians, and Nurses, Journal of Genetical Counseling 10(2) (2001), 97-119.
Walker, T.O.M. Respecting autonomy without disclosing information, Bioethics 27 (2013), 388-394.
Wetterstrand, K.A. DNA sequencing costs: data from the NHGRI Genome Sequencing Program (GSP). (URL: www.genome.gov/sequencingcosts). Accessed 2 March 2013.
Wiese, G. Gibt es ein Recht auf Nichtwissen? In: Jayme et al. (eds). Festschrift für Hubert Niederländer, Heidelberg, 1991, 475-488.
Wiesemann, C. Is there a right not to know one's sex? The ethics of "gender verification" in women's sports competition, Journal of Medical Ethics (37) (2011), 216-220.
Wiesing, U. Ethik in der Medizin, Reclam, Stuttgart, 2008.
Wilfond, B., Ross, L.F. From genetics to genomics: ethics, policy, and parental decision-making, Journal of Pediatric Psychology 34(6) (2009), 639-647.
Wilson J. To know or not to know? Genetic ignorance, autonomy and paternalism. Bioethics (19) (2005), 492-504.
Wojcicki, A. Just the facts, please. To the Editor, Nature Biotechnology 31(12) (2013), 1075-1076.
World Health Organization (WHO). Ottawa Charter for Health Promotion, 1986. First International Conference on Health Promotion, Ottawa, Canada, 17–21 November 1986. (URL: www.euro.who.int/__data/assets/pdf_file/000 4/129532/Ottawa_Charter.pdf?ua=1). Accessed 22 February 2015).
World Health Organization (WHO). Review of Ethical Issues in Medical Genetics, 2003. (URL: http://whqlibdoc.who.int/hq/2003/WHO_HGN_ETH_00. 4.pdf). Accessed 7 January 2014).
Zettler, P.J., Sherkow, J.S., Greely, H.T. 23andMe, the Food and Drug Administration, and the future of genetic testing, JAMA Internal Medicine 174(4) (2014), 493-494. doi: 10.1001/jamainternmed.2013.14706.

6

Understanding the Complexity of Regulation in an Evolving Health Technology Landscape

Andrei Famenka, Shannon Gibson, Fruzsina Molnár-Gábor

The regulation of new medical technologies presents a host of challenges, particularly in light of the rapidly evolving nature of these new interventions and the often unpredictable nature of their impact on both medical practice and, more broadly, on society as a whole. The chapter considers two different case studies of the application of different governance models to the genomic context. The first case study by Famenka looks at the expansion of newborn screening (NBS) programmes in several post-Communist Eastern European countries, and examines the policy issues and the influence of political context on the scope and content of genetic and genomic programmes. In contrast to Western European countries, where a systematic and transparent approach to genetic screening policy decisions has been established for a long time, policy-making processes in Eastern European countries still lack transparency and public participation, resulting in significant gaps between the goals of NBS programmes and what actually happens in reality. The lack of attention paid to the ethical, legal and social aspects of relevant policies might have harmful consequences for actual implementation and outcomes of NBS programmes and might undermine public trust in medical genetics and genomics. The second case study by Molnár-Gábor takes an in-depth look at the relationship between private rules and state law in the self-regulation of research around whole-genome sequencing. The scientific committee of the interdisciplinary research group EURAT in Heidelberg, Germany has adopted a Code of Conduct for Whole Genome Sequencing (2013). The increased importance of a professional norm generation stands in contrast to a deficit of knowledge of its norm qualification, especially the question of how such norm-setting can gain legal relevance, foremost according to civil liability. The further advantageous relevance of such norm-setting, for example in the timely follow-up of research development, should be compared with its potential disadvantages, such as the risk of undesirable scientification.

Introduction

The regulation of new medical technologies presents a host of challenges, particularly in light of the rapid evolution of these interventions and the often unpredictable nature of their impact on both medical practice and, more broadly, on society. As the field of genomics has advanced, genetic testing has become much more common in medical practice.[1] The emergence of many new, highly sensitive technologies is beginning to allow individuals to be screened for a range of genetic disorders and to identify particular genes or combinations of genes associated with particular health risks. Genetic testing may be used in the clinical setting for a variety of purposes: "to determine the genetic cause of a disease, confirm a suspected diagnosis, predict future illness, detect when an individual might pass a genetic mutation to his or her children, and predict response to therapy."[2] Early genetic tests detected chromosomal abnormalities and mutations in single genes, which are predominantly associated with rare, inherited disorders such as cystic fibrosis and Huntington's disease. However, more recent genetic tests can be used to detect a much wider range of conditions: "[t]here are now tests involving complex analyses of a number of genes to, for example, identify one's risk for chronic diseases such as heart disease and cancer, or to quantify a patient's risk of cancer reoccurrence."[3] With the improvements in genetic and genomic research, the rate at which new disease genes are being identified is outpacing the ability of health professionals and policy makers to assess the possible benefits and pitfalls of these technologies. Ensuring the proper regulation of new

---

[1] National Human Genome Research Institute, 2015.
[2] National Human Genome Research Institute, 2015.
[3] National Human Genome Research Institute, 2015.

genetic and genomic technologies has thus become a key issue for policy makers.

This chapter presents two different case studies that demonstrate the merits and risks of different regulatory regimes in the context of genetic and genomic technologies. Regulatory regimes can be seen as occupying a spectrum, with direct government regulation at one end and self-regulation at the other.[4] While each of these regulatory approaches has advantages and disadvantages, the successful application of each largely depends on the appropriate context and proper governance methods. At one end of the regulatory spectrum, the first case study by Famenka on the expansion of newborn screening (NBS) programmes in several post-Communist Eastern European countries, now known as countries of the Commonwealth of Independent States (CIS), explores the importance of the context in which development and implementation of government public health programmes related to human genetics and genomics take place. In contrast to Western European countries, where a systematic and transparent approach to genetic screening policy decisions has been established for a long time, policy-making processes in Eastern European countries still lack transparency and public participation, resulting in significant gaps and drawbacks in NBS programmes. Famenka examines how a lack of attention to the ethical, legal and social aspects of relevant policies can have harmful consequences for the actual implementation and outcomes of NBS programmes and lead to the erosion of trust in medical genetics and genomics.

At the opposite end of the regulatory spectrum, the second case study by Molnár-Gábor takes an in-depth look at the relationship between private rules and state law in the self-regulation of research around whole-genome sequencing. The development of standards within research in the form of guidelines and codes of conduct is of great importance in medical and healthcare systems. The increased importance of professional norm generation stands in contrast to a deficit of knowledge regarding the qualification and empirical relevance of these norms, especially for the question of how such norm-setting can become law or replace law. These questions are investigated through the example of the Code of Conduct for Whole-

---

[4] Epps, 2007.

Genome Sequencing, which was adopted in 2013 by the scientific committee of the interdisciplinary research group EURAT in Heidelberg, Germany. Such professional norm-setting complies with standards defined in international documents and is also in line with tendencies in other European jurisdictions, specifically with those in the UK, where various organisations are engaged in developing codes of conduct and best practice guidelines. Molnár-Gábor provides a detailed analysis of the significance of such norm-setting for the timely follow-up of research development, as well as potential drawbacks, such as the potential for undesirable scientification.

Towards Both Privacy and Transparency

More than 70 years ago, US Supreme Court Justice Louis Brandeis coined the phrase "sunlight is the best disinfectant" to highlight the importance of transparency in public policy. This principle of transparency is now widely regarded as an integral component of accountability, not just within government, but also within private industry. Broadly speaking, transparency refers to the need to make "relevant, timely and useful information available to the public in easy to access formats."[5] Information on key decisions and actions is made available to the public in a timely and accessible matter, including details of how and why particular decisions were made. Ultimately, transparency may lead to better decisions because decision-makers have to justify their choices based on the available evidence.

There is increasing recognition of the importance of transparency, openness and accountability in all systems of regulation no matter where they fall on the regulatory spectrum. Despite the very different contexts of the two case studies in this chapter, both highlight how the process through which standards and norms are developed has an important impact on the ultimate success of the regulatory scheme. The first case study illustrates how the imposition of standards from above, without any process of consultation or engagement, can lead to significant gaps in the regulatory framework and opposition from

---

[5] Health Canada, 2015.

within the regulated sector – gaps which may hinder the actual implementation and outcomes of NBS programmes. Famenka emphasizes the need for government to engage with the scientific community, health professionals and the public in developing frameworks for the integration of genomic knowledge into public health practice. Ideally, these frameworks should support evidence-based policy making, a transparent and coordinated community consultation process, and strong protection of individual interests and rights.

In contrast, the development of the EURAT Code of Conduct discussed in the second case study stands as an example of interdisciplinary policy development from within the regulated sector. The development of the Code included stakeholders from different disciplines involved in genome sequencing, including medical doctors, scientific researchers, economists and legal and ethical experts. This collaborative and interdisciplinary approach to policy development benefitted from its ability to draw on the expertise and experiences of the group members in addressing the normative challenges of whole genome sequencing.

Concurrent to the increasing focus on regulatory transparency, in recent decades there has been a deepening recognition of the right to self-determination in making health-related decisions and the attendant importance of privacy and confidentiality of individual health information. Many jurisdictions have been enacting new privacy legislation, or expanding the scope of existing laws, to address the privacy and confidentiality of personal health information, with some legislation introducing specific policies related to genetic information For example, in January 2011, the European Commission released a *Draft Regulation on the Protection of Individuals with Regard to the Processing of Personal Data and on the Free Movement of Such Data* (General Data Protection Regulation). Along with several other categories of data, the Draft Regulation identifies "genetic data" as a category of personal data that requires special protection.[6]

---

[6] "Genetic data" are broadly defined in the draft regulation as "all data, of whatever type, concerning the characteristics of an individual that are inherited or acquired during early prenatal development"– a definition that may encompass not

Numerous human rights documents highlight the intimate nature of genetic information and recognize a variety of rights that people have in relation to genetics. In 1997, the United Nations Educational, Scientific and Cultural Organization (UNESCO) issued the *Universal Declaration on the Human Genome and Human Rights,* the first international text on the ethics of genetic research. The Declaration aims to set universal ethical standards on human genetic research and practices in a way that balances the freedom of scientists to pursue their work with the need to protect human rights and prevent potential abuses.[7] In 2003, UNESCO followed up the 1997 declaration with the *International Declaration on Human Genetic Data,* which established principles for the collection, processing, use and storage of human genetic data.[8] Among other things, the 2003 declaration highlights the importance of free and informed consent for the collection, use and storage of genetic information and that genetic information "linked to an identifiable person should not be disclosed or made accessible to third parties, in particular, employers, insurance companies, educational institutions and the family."[9]

The requirement for informed consent may be viewed as complementary to the principles that underlie the need for transparency: in order to give valid consent, individuals must be informed of the reasons for the collection, use and storage of genetic information, as well as the potential risks and consequences of such activities.[10] As explored in the first case study, the absence of an informed consent requirement for newborn screening programmes in

---

only genetic information, but also family medical histories and related health information: European Commission 2012, 42.

[7] UNESCO, 1997. Article 10 of the Declaration states that "[n]o research or research application concerning the human genome, in particular in the fields of biology, genetics and medicine, should prevail over respect for the human rights, fundamental freedoms and human dignity of individuals or, where applicable, of groups of people."

[8] UNESCO, 2003. However, as declarations, the two above instruments are not in themselves legally binding. Yet the principles and rights that underlie many of the provisions set out in the declarations are based on human rights standards set out in other international instruments that are legally binding, such as the United Nations *International Covenant on Civil and Political Rights.*

[9] UNESCO, 2003, Art. 14(b).

[10] UNESCO, 2003, Art. 6(d).

CIS countries has resulted in poor communication with parents and limited awareness by the general public of the goals and methods of the programmes. Without adequate information about NBS, neither parents nor the public can make informed decisions about these programmes. Famenka notes that this problem is exacerbated by the lack of data collection on NBS programmes in CIS countries, leading to an absence of data on long-term outcomes and uncertainty about the efficacy of treatment for a number of diseases.

The rights and principles enumerated in the two UNESCO documents, as well as in numerous other similar instruments, are now widely regarded as foundational in genetic and genomic research. Indeed, the EURAT Code described in the second case study incorporates many basic principles of bioethics, including good scientific practice, privacy protection, risk management and the requirement for informed consent. The importance of transparency is also reflected in the introduction to the Code, which states that the scientific community has an "obligation to inform society about the methods of its research, its aims, and its results, as well as the associated risks." Molnár-Gábor also notes that the EURAT Code is intended to establish a culture of trust that will ensure both the willingness of participants to donate their genomic data and the support of such research by the public sector.

Increasing Complexity in the Genomic Era

Genetic testing is performed for a wide range of different purposes and in many different contexts. As such, the regulatory requirements vary widely depending on the test being performed and the purpose of testing. Soini notes that "trying to make policies and govern the use of genetic information is a challenging mission since the basic question is what is [it] we are trying to regulate and why and what are we trying to protect."[11] Moreover, as Famenka discusses in the first case study, the governance models that have traditionally been employed for population screening programmes such as NBS may not be well adapted to the rapidly advancing technologies that are

[11] Soini, 2012, 144.

being developed in the "genomics era." This is due in large part to conceptual and practical difference between genetic and genomic information; as the technology has evolved, new ethical, social and legal challenges have come to light.

As noted above, transparency and consultation are important foundational principles in the governance of new health technologies. This effect may be particularly pronounced in the context of genetics and genomics because of the (at least perceived) sensitivity and intimacy of such information, and the huge amount of information that can be derived from a single test. The term "genetic exceptionalism" is often used to describe the argument that genetic information is different from other types of health information and, therefore, deserves a higher level of protection for issues of privacy, confidentiality and informed content. One argument that is commonly put forward to support the exceptional nature of genetic information is its potential to predict the probability that a person will develop a particular disease or condition in the future. As Green and Botkin point out, although both genetic and non-genetic tests "can identify risk factors for future illness, detection of highly penetrant genetic mutations may indicate a substantially higher risk than abnormalities discovered by nongenetic tests."[12] They also note that society often views genetic information as being more intimate and "central to our core being" than other types of personal health information which, rightly or wrongly, may contribute to the public perception that genetic information should be accorded a higher level of protection.

Another commonly cited distinguishing factor of genetic information is that due to its heritable nature, genetic information also has an impact on family members, and even on the broader community. Genetic tests "identify predispositions that are exclusively transmitted vertically (from parent to child), while non-genetic tests identify predispositions transmitted in a variety of ways (exposure to common environmental risk factors or person-to-person contact)."[13] As such, the decision of one family member to be tested for a heritable disease or risk factor may lead to a definitive diagnosis

---

[12] Green/Botkin, 2003, 572.
[13] Green/Botkin, 2003, 572.

or risk factor identification for a family member who otherwise might have preferred not to know such results.

The high degree of concern around the sensitivity of genetic information is reflected in the types of information legislation and policies that have been adopted in some jurisdictions to grant special protection of genetic information. One of the first legally binding international documents to include prohibitions against genetic discrimination was the 1999 *European Convention on Human Rights and Biomedicine*, which sets out a "series of principles and prohibitions against the misuse of biological and medical advances,"[14] including prohibiting genetic discrimination (Article 11), and restricting the use of predictive genetic testing to health care and scientific research purposes (Article 12).[15] Many jurisdictions have also enacted specific legislation to prohibit genetic discrimination. For example, in 2008, the US government passed the *Genetic Information Nondiscrimination Act* (GINA), which prohibits health insurers or employers from discriminating against individuals based on genetic information.[16]

While a detailed analysis of the justification for genetic exceptionalism is beyond the scope of this chapter, overall it is clear that genetic information has the potential to reveal significant information about an individual's current and future health status, and further, this information may have implication for both family members and the broader community. Advances in predictive genetic testing have tended to exacerbate concerns around the potential for unfair or restrictive treatment of individuals based on genetic information.[17]

Whole genome sequencing stands as a prime example of both the technological progress and governance challenges that are arising in the genomic era. As will be discussed in the second case study, there are a number of important features that differentiate whole genome

---

[14] Council of Europe, 1999.

[15] Council of Europe, 1999.

[16] Genetic Information Nondiscrimination Act of 2008. In particular, health insurers may not deny coverage or charge higher premiums to otherwise healthy individuals based sole on genetic predisposition to develop a disease in the future, and employers may not use an individual's genetic information in making hiring, firing, job placement, or promotion decisions.

[17] Green/Botkin 2003.

sequencing from conventional genetic testing that may exacerbate some of the concerns around genetic exceptionalism. For example, Molnár-Gábor notes that whole genome sequencing reveals much more information than a single genetic test, which may increase the risk of re-identification from even very few DNA sequences. In addition, whole genome sequencing may reveal findings unrelated to the patients' initial medical question (primary findings), but which may nonetheless be of medical value or utility to the ordering physician and the patient (secondary findings). Such secondary findings, as well as additional findings without direct medical value or utility, can complicate the informed consent process.

From Direct Regulation to Self-Regulation

In the first case study, Famenka notes how the new complexities that are arising in the genomic era create the need for new guidelines and codes of conduct to govern the complex legal, ethical and social issues around genetic and genomic technologies. Moreover, the introduction to the EURAT Code, discussed in the second case study, specifically states that in response to increasingly complex research methods, the scientific community has begun developing its own guidelines and codes to codify good scientific practice. This hints at a trend towards self-regulation where the sector or profession, rather than the government, plays an increasingly active role in regulation.

Proponents of direct government regulation often argue that it offers a number of advantages "in terms of visibility, credibility, accountability, compulsory application to all… greater likelihood of rigorous standards being developed, cost spreading… and availability of a range of sanctions."[18] Unfortunately, many of these purported benefits fail to play out in practice. Indeed, the first case study highlights some of the problems that may arise in the direct regulation of genetic screening programmes. For example, Famenka notes that there can be significant gaps between official policy and what actually happens in reality where government programmes are

[18] Epps, 2007, 78.

not properly implemented or managed. Further, due to the complex and rapidly changing nature of genetic and genomic technologies, direct government regulation is often a poor choice since the legislative process may be slow and cumbersome, leading to inflexible rules that are difficult to update.

In contrast, in the second case study, Molnár-Gábor discusses how private self-regulation tends to be more timely and responsive to scientific development than direct government regulation. Moreover, the regulated community often has more expertise and technical knowledge of practices in their sector than government regulators.[19] Self-regulation may also allow for greater flexibility than direct government oversight as a result of less formal rules and processes, and further, may reduce the adversarial stance between government and the regulated sector.[20] However, the second case study also notes that while self-regulation has the advantage of disciplinary appropriateness, it does create the risk of over-scientification since codes and guidelines are typically developed by experts.

Finally, Molnár-Gábor notes that as the science becomes more complex, the interpretation of test results requires increasing genetic expertise; medical doctors may lack the knowledge to fully understand the test results, let alone its relevance to patients. Moreover, whole genome sequencing often involves numerous clinics and research institutions, which can further complicate governance. The second case study also notes that there is a trend towards international collaboration in genetic and genomic research, further confounding regulatory efforts. The trend towards international genetic and genomic research requires a corresponding evolution in the regulatory environment.

Conclusion

Overall, the governance of new genetic and genomic technologies is a complex and ever-changing landscape. As the case studies in this chapter demonstrate, the regulation of new health technologies is no

---

[19] Ibid.
[20] Epps, 2007, 80.

simple task, and while the contexts of regulation may vary significantly, the need for transparent and accountable regulatory processes is a common theme. Moreover, the rapidly evolving nature of new genetic and genomic technologies requires that regulatory processes be capable of evolving alongside the technology; the old models of direct government regulation that have dominated traditional genetic screening programmes may not be well suited for the more complex technologies of the genomic era, such as whole genome sequencing. Finally, governance mechanisms and systems must be capable of adapting to keep up with the evolution of the technology. In this context, the shortcomings of direct government regulation and the advantages of more flexible systems of self-regulation are becoming apparent.

Case Study 1 – In Need of Transparency: The Case of Genetic Screening Policies in Post-Communist Eastern European Countries

Andrei Famenka

Introduction

In the past decade, advances in human genetic research have substantially enriched our knowledge of the role genes play in health and disease.[21] During this period, the costs of laboratory technologies have decreased dramatically, thus making it possible for genetic tests to be applied widely in health care.[22] Also, with the sequencing of the entire human genome, new opportunities have emerged for medical professionals to use this specific knowledge more precisely and effectively in their efforts to promote the health and wellbeing of individual patients and the public.[23] [24] Translating genomic knowledge and assuring its appropriate use in public health are therefore becoming key issues for both health professionals and policy-makers.[25] [26] However, the mechanisms of such a translation appear to be complex and controversial, as they are likely to involve a range of issues of ethical, legal and social origin, raised by the integration of new genomic knowledge into public health programmes.[27] Conditions and frameworks under which genomic knowledge can be put into the policy and practice of public health therefore need to be designed carefully and defined clearly, in order to take all the relevant interests into account with the aim of preventing possible misunderstandings and abuses.[28]

---

[21] Guttmacher/Collins, 2002.
[22] Peltonen/McKusick, 2001.
[23] Venter et al., 2001.
[24] Collins/McKusick, 2001.
[25] Yoon, 2001.
[26] Ojha/Thertulien, 2005.
[27] Clayton, 2003.
[28] Brand, 2008.

Genetic screening is often considered as a paradigm case for integration of genetics into public health.[29] However, the model of genetic services, traditionally employed by population screening programmes before the coming of "genomics era", might not be fully appropriate for dealing with genomic data due to conceptual and practical difference between genetic and genomic information.[30] What transformations should existing genetic screening programmes undertake in order to be applicable in the new setting of the "genomics era"? Taking the case of the expansion of newborn screening (NBS) programmes in several post-Communist Eastern European countries, now known as countries of the Commonwealth of Independent States (CIS), I am going to explore the importance of a context in which the development and implementation of public health programmes related to human genetics and genomics take place. As the lack of transparency or public participation is considered to be a distinguishing characteristic of policy processes in the majority of CIS countries,[31] their implications for the development of newborn genetic screening programmes in these countries are discussed. Examples of significant gaps and drawbacks in existing genetic screening programmes are provided in order to demonstrate the harmful effects that inadequate policies can have for the actual implementation and outcomes of NBS programmes. I also emphasize the need for the concerted efforts of government, the scientific community, health professionals and the public in establishing socially just frameworks for the integration of genomic knowledge into public health practice. Finally, I draw attention to the harmful consequences of dysfunctional NBS programmes which, being introduced with little attention to the ethical, legal and social aspects of relevant policies, could lead to the erosion of the general public's trust in medical genetics and genomics.

---

[29] Khoury et al., 2003; Nuffield Council, 2006.
[30] Khoury, 2003.
[31] Dryzek/Holmes, 2002.

Newborn Screening Programmes

Newborn screening for genetic disorders has become a paradigm case of the integration of genetics in public health, due to its widespread use and long history of implementation.[32] In the form of public health programmes, NBS was first introduced in the late 1960s to early 1970s in the majority of Western countries, and was then adopted by Eastern European countries as well. From the very beginning, NBS programmes were focused on a few, relatively well understood disorders, for which effective treatments were available. The basic set of conditions for which all the newborns in most countries of the world were supposed to be screened initially consisted of phenylketonuria (PKU) and hypothyroidism (HT). The programmes targeted to these disorders often serve as a foundation for other types of genetic screening. Indeed, these programmes have proved to be quite successful, both in terms of their efficiency in reducing the burden of serious morbidity and their ability to cover almost the entire population of newborns. Historically, largely because of the significant benefits and small burden associated with these programmes, they were perceived as mostly unproblematic from the ethical and social point of view. In the early years of their implementation, PKU and HT newborn screening programmes fully met the criteria of acceptance of population screening programmes, developed by Wilson and Jungner and endorsed by WHO.[33] These criteria stressed the importance of a given condition to public health, the availability of a screening test, the availability of treatment, and cost effectiveness.

PKU and HT are indeed serious conditions, although quite rare (the prevalence of PKU among newborns is about 1 in 25000), and without timely and adequate treatment they will inevitably cause severe health impairment. However, the tests for PKU and HT are cheap and easy to perform, and can be undertaken in the first days of life by taking several drops of baby's blood for analysis.[34] Once the disease is revealed, affected infants are supposed to be given a special

---

[32] Khoury et al., 2003; Nuffield Council, 2006.
[33] Wilson/Jungner, 1968.
[34] American Academy of Pediatrics, 1989.

diet (in the case of PKU) or hormone therapy (HT) that can prevent health damage. At the time the first NBS programmes were launched, the considerable benefits they brought for affected children as well as for the public were thought to be sufficient justification for implementing them on mandatory basis. Along with the development of newborn genetic screening technologies, e.g. using of filter paper for processing and storage of dried blood samples, there was a parallel development of relevant infrastructure (procurement and storage facilities, logistics services and diagnostic laboratories). Contemporary justifications for NBS programmes rest on the state's authority given the presumed benefits to children, and the efficiency of these programmes to identify affected babies and provide them with adequate treatment.

However, with the sequencing of the human genome and emergence of some new, highly sensitive and rather cheap technologies, which enable one test to screen for a range of genetic disorders and also make it possible to identify particular genes or their combination, new ethical, social and legal challenges have arisen for these public health programmes. It seems that the traditional conceptual foundations of NBS programmes do not accommodate the potential of these new technological advances very well. With the improvements in genetics and genomics, the rate at which new disease genes are being identified is outpacing the ability of health professionals and policy-makers to assess the possible benefits and pitfalls of expanding newborn genetic screening programmes. These challenges are even more severe for the post-Communist Eastern European countries, where a wide variety of different factors of transition continue to make considerable impact on how genetic screening programmes are developed and implemented.

Expansion of Neonatal Screening Programmes in CIS Countries

Within the past decade, there has been a visible trend towards expanding the scope and nature of neonatal genetic screening programmes in CIS countries. In Russia, the number of genetic conditions included in the mandatory screening programme has risen threefold over the past few years, and in Ukraine and Belarus, proposals have been made to expand existing programmes significantly.[35] However, this rapid expansion has led to controversy over the appropriate targets of screening, as many of the conditions that have recently been included in the standard panel are relatively rare and poorly understood, and for most of them an effective treatment is not yet available. Also, the introduced methods are different from the traditional methods of screening for PKU and HT, which are indirect and do not identify the genes involved. In contrast, new methods are able to identify both affected individuals and those who are carriers of a gene for a recessively inherited disorder. The finding of a carrier has no health implications for the child, but may become important for the family and for that child in a variety of ways: knowledge of carrier status can lead to testing of the parents and family members, concern may arise about possibly affected future siblings if both parents are carriers, there is a possibility that screening might reveal that the male partner is not the biological father, concern may arise about the child's future reproductive choices, and there may be anxiety (although unjustified) about the health of the carrier newborn. However, it seems that the existing public health services that make up the infrastructure of NBS programmes are not sufficiently prepared for working with carriers, as these programmes were designed only for identifying and treating affected children. The lack of counselling services in these programmes, the absence of professional guidelines, and the lack of professional and public education mean that health professionals involved in NBS programmes may be ill-equipped to respond to the challenges of carrier identification. Controversy has therefore arisen over whether the expansion of newborn screening programmes is ethically justified when benefits are not well established and risks are not entirely clarified. Do the post-Communist Eastern European countries have sufficient capacity to meet these challenges and react adequately to the emerging problems? What changes should be introduced to relevant policies in order to make these programmes genuinely beneficial for those who need help and at the same time not cause harm to any of those involved?

---

[35] Ministry of Health of the Republic of Belarus, 2009; Ministry of Health and Social Development of the Russian Federation, 2006; Charchenko, 2012.

Officially, public policies in CIS countries proclaim laudable goals, such as preventing serious morbidity and promoting the health of the population. However, there can be serious gaps between official public policies and what actually happens in reality, as programmes for the promotion of policy goals can be inadequately constructed, implemented or managed. In this regard, newborn screening programmes in CIS countries can serve as an example of inadequate policy-making, which seems to prioritize the promotion of a biotechnology sector and economic growth, while paying much less attention to developing appropriate infrastructure, ensuring equal access to health services and promoting justice in benefit sharing. Lack of transparency or public participation in the decision-making process exacerbates the gaps in policies even further.

In contrast, in the majority of Western countries, public health agencies use the approach of combining the scientific knowledge base with extensive public consultations in decision making and the formulation of public health policy. In any formulation of sound public health genetic policy, relevant ethical, legal and social issues are considered before genetic tests are applied in population-based prevention programmes. In the United Kingdom, the government established the Human Genetics Advisory Commission in 1996 to review scientific progress, report on issues that have "social, ethical and/or economic consequences, for example in relation to public health, insurance, patents and employment" and to "advise on ways to build public confidence in, and understanding of, the new genetics." [36] The Nuffield Council on Bioethics in the UK has produced a report and a supplement, which summarize the ethical issues of genetic testing and promote discussion of the development of future policy and practice.[37]

In the United States, the Secretary's Advisory Committee on Genetic Testing was established in 1999, to advise the Department of Health and Human Services on the medical, scientific, ethical, legal and social issues raised by the development and use of genetic tests.[38] In the same year, the American Academy of Pediatrics and the Health

---

[36] Human Genetics Advisory Commission, 1996.
[37] Nuffield Council on Bioethics, 2006.
[38] Secretary's Advisory Committee on Genetic Testing, 1999.

Resources and Services Administration convened the Newborn Screening Task Force to address the lack of consistency in the disorders included in screening programmes and the testing methods used in the various states.[39] The group concluded that there should be a national consensus on the diseases tested for by state programmes of newborn screening. The American Academy of Pediatrics, American College of Medical Genetics, Health Resources and Services Administration, Centers for Disease Control and Prevention, and other groups worked together to create a national agenda for newborn screening.

The analysis of NBS programmes operating in CIS countries shows that they have been created without public discussion and have a range of ethical, legal and social implications. For example, the major problem with NBS programmes in CIS countries is the lack of support for long-term follow-up and care. The major part of the resources for NBS goes to laboratory services and short-term follow-up.[40] [41] In fact, NBS programmes follow affected children only long enough to make sure they have received a definitive diagnosis and obtained initial treatment services. Families may be guided to where they can receive expert treatment, but treatment is not provided by the screening programme. In these circumstances, NBS programmes are only as effective as the larger health care systems in providing services to affected children. Here, it should be noted that health care systems in CIS countries constantly experience severe budgetary constraints; provision of health care services in these countries is unstable and depends deeply on a combination of different political and economic factors.[42]

Although health care systems in CIS countries are funded from taxation and are supposed to offer universal access to health care for their citizens, only those services that are included in the list of "essential services" are free at the point of use. Yet treatment for some conditions that are included, or are intended to be included, in the expanded NBS programmes, is excluded from the list. In Belarus,

---

[39] Newborn Screening Task Force, 2000.

[40] Ministry of Health of the Republic of Belarus, 2009.

[41] Ministry of Health and Social Development of the Russian Federation, 2006.

[42] Rechel/McKee, 2009.

the treatment for cystic fibrosis, which is expensive and must be maintained throughout a lifetime, is not provided for free by the state health care facilities. Families of affected babies have to seek funds to cover treatment expenses from charity organizations and donations, but such efforts do not guarantee that treatment will always be provided in time. Even for conditions that are well known, there are occasions when treatment is not available for financial reasons. There have been several quite long interruptions in the supply of medical food for PKU in Ukraine, which caused an extensive media scandal and resulted in furious court battles initiated by the families of affected children against the health authorities.[43] These examples clearly demonstrate that the value of a system for early detection of rare genetic conditions is substantially reduced by the failure to meet the special needs of the children with disorders. In other words, when sustainable long-term care is absent, NBS provides little benefit to affected children. At the same time, it is quite clear that at least in the near future, the limited resources available to state public health programmes in CIS countries will not be expanded to cover treatment services for conditions detected through NBS.

In CIS countries, parental education and active engagement in NBS decision-making remains a major challenge. Despite the significant expansion in the scope and complexity of these programmes, there have been no attempts to improve the education of parents about NBS. However, with the capacities of the new NBS tests to detect genes for recessively inherited disorders or/and genetic abnormalities that confer an increased risk to the individual rather than the certainty of developing the disease, the scope and nature of the information to be provided to parents become more complex and harder to interpret than information provided within the frameworks of traditional NBS programmes. The complex nature of this new information and its potential strong ethical, legal and psychosocial implications make the presence of adequate counselling and education even more important. As the tools for evaluating and communicating complex genomic information, even for longstanding and well-established NBS programmes, like those of PKU and HT, are largely underdeveloped in CIS countries, misunderstanding the

[43] Charchenko, 2012.

purposes of the tests by new parents is not a rare phenomenon. The lack of understanding and efficient collaboration between health care professionals and parents can negatively impact on the provision of initial screening, evaluation and follow-up services for affected children, as these children may be harmed through slow or inappropriate responses by parents (or clinicians) to screening test results. Therefore, policy-makers should understand that adequate education of patients is an important component of a comprehensive approach to screening. Currently, though, the information needs of parents in CIS countries are only met through passive mechanisms such as providing brochures, as there is no informed consent requirement for the NBS in the law. Most often, the distribution of educational materials occurs during the immediate postnatal period, a time when new parents need to address other priorities in newborn and maternal care.

Another problem for NBS programmes in CIS countries is the lack of data collected on affected children and their families. As state programmes typically do not collect data on children following a definitive diagnosis and referral for initial care, they do not have data on the long-term outcomes of these children.[44] [45] With the expansion of the standard panel of NBS programmes, and given the rare nature of these conditions, it is becoming more and more difficult to collect data on clinical outcomes, thus leaving uncertainty about the efficacy of treatment for a number of diseases. Developing a long-term follow-up system is complex and expensive. However, as resources are nevertheless used to add more conditions to the screening panel, it is appropriate to ask whether resources should instead be devoted to ensuring that affected children are given adequate care and that data are collected to evaluate the efficacy of evolving treatment strategies.

An important characteristic of state-based newborn screening programmes in the post-Communist Eastern European countries is that they are mandatory.[46] However, while the states officially enable

---

[44] Ministry of Health and Social Development of the Russian Federation, 2006.

[45] Ministry of Health of the Republic of Belarus, 2009.

[46] Ministry of Health and Social Development of the Russian Federation, 2006; Ministry of Health of the Republic of Belarus, 2009.

parents to opt out of NBS for religious or moral reasons, parents are not adequately informed about the choice. The law in these countries does not require signed informed consent for the conduct of NBS. As a result, the lack of a consent requirement leads to poor communication about NBS with parents and limited awareness in the general public about the goals and methods of the programmes. It appears that health authorities can be reticent about providing information on NBS to parents, in part due to fear that parents may decline screening and a disorder may be missed. However, the strategy has not proved to be successful, as more and more parents now opt out of NBS altogether because of misinformation and misunderstanding about the programmes. As a result, health departments have found themselves caught between the risk of parents declining screening through an opt-out system, and the risk of parents declining screening because of loss of trust in providers of the programmes. Parental education on NBS is therefore an issue of social justice, as without adequate information about NBS parents cannot make informed decisions and may therefore create health risks for their children.

Conclusion

The case of expansion of NBS programmes in CIS countries illustrates a broad spectrum of issues of ethical, social and legal origin, which emerge when attempts are made to integrate the recent advances of genomics into public health policy and practice. The case also illustrates that ignoring these challenges and paying little attention to the issues of transparency and community participation in the process of decision making can have harmful effects on outcomes of the policy. The review of contemporary NBS policies and corresponding public health interventions in CIS countries shows that the absence of comprehensive, evidence-based policy aimed at integrating new genetic and genomic knowledge into public health undermines social justice in terms of ensuring equal access to the benefits of these programmes. In the context of transition, the expansion of NBS programmes has generated only limited professional and public awareness, and this systematic lack of transparency is beginning to harm programmes through an erosion of public trust. The case of expanding of NBS programmes in CIS countries suggests that the effective translation of genetic and genome-based knowledge into public health requires concerted efforts from government, the scientific community, health professionals and the public, aiming at ensuring favourable conditions and support for evidence-based policy making, transparent and coordinated community consultation process, and strong protection of individual interests and freedom.

Case Study 2 – Self-Regulation in Research: The EURAT Code of Conduct for Whole Genome Sequencing

Fruzsina Molnár-Gábor

Introduction: Whole Genome Sequencing (WGS)[47]

The Evolving Technology

DNA – the biomolecule that carries genetic information – can be isolated from a simple blood sample. After fragmentation and reproduction of the genome, different sequencing methods allow the parts produced to be compared with healthy sequences based on their length and biological patterns.[48] Conclusions can then be drawn about the genetic background of predisposition to certain diseases.[49] Comprehensive next-generation sequencing (NGS) technologies extend this process and enable rapid sequencing in long fragments of base pairs throughout the genome.[50] These high-throughput analytical methods show details of the structure and function of human genetic material, which make it clear not only that individual genomes vary to a great extent, but also that this has a crucial impact on the origin, the predisposition towards and the severity of illnesses.[51]

Research efforts are currently directed towards the application of NGS in medical care.[52] Comprehensive *research projects* allow

---

47 This introduction is a revised version of the introduction to the article: Molnár-Gábor, 2014.

48 Murken/Grimm/Holinski-Feder, 2006, 98.

49 Scherrer, 2011, 7.

50 Heinig/Kalle v., 2012; A clear explanation is also available on the website of Illumina, http://www.illumina.com (Accessed 14 Dec 2014).

51 Compare the description: http://www.uni-heidelberg.de/totalsequenzierung/informations/issues.html (Accessed 14 Dec 2014).

52 Some countries have already started to integrate NGS in the clinical diagnosis of particular illnesses. (Presidential Commission for the Study of Bioethical Issues, 2012, 4.) In addition to the USA and Canada, other countries such as China, South Korea, India and Mexico are also involved. (See http://icgc.org/ [accessed 13 Dec 2014].) Numerous European countries have also taken initiatives. In the UK, whole genome sequencing being introduced into clinical practice. (Cressey, 2012.) In

sequencing technologies to be improved, leading to a significant reduction in costs.[53] This makes the technology available to a continuously expanding user group; private companies now also offer genetic profiling.[54] Most importantly, the technology is being used increasingly in *diagnostic contexts*.[55] When associated with patient data, the identification of disease-related alterations in the genome allows a better and, ideally, individualized understanding of the conventional findings, and might succeed in optimizing therapy for particular groups of patients.

The first step in applying NGS technologies in medical diagnostics takes place on a *translational level*.[56] In this context, NGS technologies such as WGS present new challenges compared to conventional medical or even conventional genetic analyses.

The special attributes of whole genome sequencing as a medical intervention and their influence on the status and the role of scientific researchers

The most important features of WGS, which allow us to differentiate it from conventional medical interventions, are the additional findings gained, its informational intervention character accompanied by the possibility of re-identification, and its practice in an international context.

First, decoding large stretches of DNA makes it possible to discover a very high number of *additional findings*. These findings do not provide accurate information about the patients' initial medical

---

Germany leading clinicians are planning to be able to sequence every cancer patient soon. (Siegmund-Schultze, 2012). See also Norway: Callaway, 2012.

[53] Hayden, 2014.

[54] 23andMe Inc., http://www.23andme.com/; bio.logis GmbH, http://www.biologis.com/; AITbiotech, http://www.aitbiotech.com/. Accessed 13 December 2014.

[55] Borrell, 2010; Kohlmann, 2012; Green/Berg/Wayne, 2013; American College of Medical Genetics and Genomics. ACMG Updates Recommendation on "Opt Out" for Genome Sequencing Return of Results, 2014 See: https://www.acmg.net/docs/Release_ACMGUpdatesRecommendations_final.pdf [Accessed 6 August 2015].

[56] Translational medicine means the direct transfer ("translation") of laboratory knowledge into clinical settings. See: Rosenberg, 2003; Sung/Crowley/Genel et al., 2003.

issue (primary findings) that made them participate in a specific translational study, and usually also go beyond secondary findings that are unrelated to the reason why the sequencing was undertaken but that may be of medical value or utility to the physician and the patient. Additional findings may provide predictive information about different dispositions. Even though they can help with early detection of diseases, they often do not indicate concrete and reliable preventive measures or treatment securely based on medical and scientific causality. The number of such findings and their interpretation is exposed to continuously changing factors due to developments in science and technology, and they depend strongly on the bioinformatic filter applied.[57] Although additional information can be expected, its volume and predictive meaning may stay incidental, depending on the filter.

Second, during the period of investigation, genetic analysis allows a continuous flow of information. The small physical impact of blood sampling cannot be compared to the actual implication of the intervention, which is the possibility of obtaining sensitive information dynamically. As the patient's stored DNA sequence can be accessed over a longer period of time, genetic tests can be carried out "*in silicio*".[58] Genome sequencing can thus be described not as single or selective, but as a continuous, successive *informational intervention*.[59]

In addition to this, because the DNA sequence itself is a truly unique identifier of each human being, progress in genetic technologies has led to the possibility of identifying individuals even when very little information is available. Although the coding of genetic data meets the most exacting requirements, we have to be aware of the fact that very few and very short DNA sequences would be sufficient to allow the unambiguous correlation between genetic material and a person. Furthermore, although the genetic data are coded they can be traced back to a person if identifiable reference material is found elsewhere.[60] If re-identification is possible using

---

[57] Sandroff, 2010, 24-25.
[58] Rehmann-Sutter/Bobbert/Dölling, 2012.
[59] Molnár-Gábor/Weiland, 2014.
[60] German Ethics Council, 2010.

only very few DNA sequences, genome wide analysis will only increase the challenge.[61] This presents a problem of "dual use", in this case the misuse of personal data by others.

Third, questions about the integration of sequencing in translational medical care are pressing, especially because most of the leading research consortia and institutions involved are set up on an extremely international level.[62] The willingness to involve a maximum number of patients in translational studies also indicates further *internationalization*.

Each of these characteristics has a crucial impact on the status and the role of scientific researchers involved in the sequencing process.

The *interpretation* of the genome sequence demands molecular genetics experts. The genetic data set derived from a sample is usually stored and then sent to a molecular genetics laboratory for analysis and validation. Extensive knowledge of all possible findings might not be available at the time of consent, and even if it is available, it might be beyond the expertise of the medical doctor. As a general rule, skilled medical doctors' understanding will not be sufficient to evaluate or to validate the *primary and secondary results* of sequencing, or to grasp their relevance for and impact on the patient. In some cases scientists may incidentally or intentionally discover *additional findings* that might also provide predictive research results.

Secondary and additional findings influence *the informed consent process*, which is the legal basis for applying WGS as a diagnostic tool. Timing and content-related difficulties in providing information on such findings complicate the process of informing the patient, and rapid changes in research benchmarks are a further obstacle. Obtaining these findings within the limits of an *informational intervention* influences the *doctor-patient relationship*. On the one hand, during this process the relationship between medical doctors and patients becomes continuous instead of a point source exchange, so

---

[61] International Bioethics Committee, 2014.

[62] The ICGC (International Cancer Genome Consortium) is a significant example, which has already announced that it has 77 cancer projects involving nearly 20 different countries, encompassing almost 20 different ideas on patients' rights and data protection. The projects of this consortium use sequenced genetic data of tumour patients. See: http://www.icgc.org. Accessed 14 December 2014.

that the need for additional consulting work increases. On the other hand, their bipolar relationship becomes looser, since non-physician scientists generally have to become involved in the interpretation of the findings.[63] The interpretation of risk information provided from outside the relationship might shift the balance within the doctor-patient relationship.

It is not clear whether researchers are obliged to notify the responsible physician of all findings including additional findings identified during the sequencing process. It is also uncertain whether researchers can be obliged to engage in the active search for additional findings beyond the specified context of a sequencing request. The opportunity to re-identify genetic information so easily goes hand in hand with an increased possibility for scientific researchers to interfere continuously in the patients' (and to a certain extent also in his or her relatives') rights.

Whole genome sequencing usually involves a number of different clinics and research institutes. Since translational studies are not limited to particular institutions but are a part of international research projects, the data collected from sequencing genomes are entered into *international databanks* together with non-genetic patient data in a coded form. Because of the increasingly international character of these projects, concerns about privacy need to be sensitive to the range of cultural issues and should aim to secure genetic information.[64] Non-physician scientists who are working with genetic data are especially affected by the challenges of data privacy and confidentiality.

Even at this first glance it is clear that new forms of responsibilities are emerging for researchers involved in genome sequencing. They do not yet have an established code of professional conduct, comparable to those of medical doctors, which motivates them to determine their responsibilities and obligations when involved in genome sequencing. Clear identification of the specific competencies throughout the complex workflows of whole genome sequencing might protect and benefit both the non-physician scientist and the patient.

---

[63] Wollenschläger, 2013.

[64] Laurie/Mallia/Frenkel et al., 2010, 319.

II. Project EURAT: Code for non-physician scientists and personnel involved in whole genome sequencing, particularly of patient genomes[65]

The EURAT Code of Conduct is a result of the work of an interdisciplinary group of outstanding representatives of different disciplines involved in genome sequencing including medical doctors, scientific researchers, and experts in applied ethics, law and economics. The idea of establishing such a code of conduct came out of their cooperation while addressing the normative challenges of WGS as defined within the aims of the EURAT project, as well as out of the personal experiences of the members of the group in their own scientific practice. There was no explicit mandate, not even from the Marsilius-College of the University of Heidelberg in Germany, which funded the project for 3 years.

> In recent years progressively more complicated research methods and results have prompted emphatic admonition that the practice of scientific inquiry should be guided by basic ethical principles. The sciences are answering this call by autonomously developing their own codes that concretize good scientific practice. But additional steps must be taken to cover the remaining and multiplying issues. For instance, there is science's obligation to inform society about the methods of its research, its aims, and its results, as well as the associated risks. For all research on human subjects there is an additional obligation vis-à-vis the patients or persons who are the object of a study. This obligation also pertains to the persons with whom researchers do not come into immediate contact, hence the need for a code for non-physician scientists. An equivalent to the obligation physicians have towards their patients could be a form of self-commitment for these researchers comparable to the Hippocratic Oath.[66]

The scientific committee of EURAT explained the adoption of the Code of Conduct with these guiding words. In addition to protecting and benefiting researchers and patients, the Code is intended to help

---

[65] EURAT project, 2013.
[66] Ibid, 23 et seq., 23 para. 1.

establish a culture of trust in order to ensure that people are willing to donate their genomic data and that the public sector is willing to provide continued support.[67]

The Code applies to all non-physician scientists and personnel who are involved in the study and analysis of patient genomes. In particular, researchers involved in life sciences research or in "preliminary diagnostic" analyses need guidelines that specify their responsibilities compared to researchers who are entrusted with evaluating patient genomes in clinical diagnostics, since the latter are more closely bound to the canon of rights and responsibilities for physicians in the clinic.[68]

The Code is divided into two parts; part one contains basic ethical principles, and part two defines practical guidelines. In addition to the basic principles of bioethics such as respect for persons, patient autonomy, beneficence and non-maleficence, non-discrimination and justice, part one also reaffirms specific principles for genetics. These are good scientific practice, privacy protection, the protection of future generations, public benefit, and the prohibition of financial gain from the genome in its natural state.[69] The guidelines in part two specify best practice and conduct regarding different procedures for whole genome analysis. These are risk management, patients' consent and data protection, including concerns about the collection and use, security, sharing and access to databases, documentation and the publication of results. Subsequently, the guidelines specify how to deal with research findings and how to guarantee the patients' right to withdraw their consent.[70]

The establishment of codes for conducting whole genome sequencing is in line with trends in other European countries such as the United Kingdom, where private entities have recently published extensive guidelines in order to push genome sequencing into clinical practice in the very near future.[71]

---

[67] Ibid, 30 para. 8.

[68] Ibid, 23-24 para. 2.

[69] Ibid, 12 et seq.

[70] Ibid, 14 et seq.

[71] Compare the reports of the PHG Foundation in Cambridge, UK, especially the report 2011, with policy recommendations, p. 153 et seq.

Evaluation of the EURAT Code of Conduct

Norm Qualification

As already noted, the EURAT Code of Conduct embodies professional standard-setting in the field of translational medicine and is not legally binding. However, it represents an independent regime of norms, which specify the ethical and practical benchmarks for conducting whole genome sequencing in patients' genetic material. The joint declaration of ethical principles and guidelines highlighting routes for decision-making and corridors of action when conducting such sequencing is complemented by the reflection of the status of the code itself regarding liability, its binding force and implementation. However, in contrast to other relevant codes of conducts for researchers – such as the Guidelines and Rules on a Responsible Approach to Freedom of Research and Research Risks of the Max-Planck-Society – it misses the detailed declaration or the reaffirmation of the relevant legal principles in this field.[72] It only states that each individual researcher is solely responsible for complying with existing legal regulations and that it is incumbent upon the individual researcher to ascertain which legal regulations are applicable to their activities and to uphold them within their sphere of competence. It also points out that primary investigators and directors of projects, departments and institutions also bear responsibility within the legal framework of vicarious liability for the conditions and practices within the entire organization over which they preside.[73]

It is obvious that self-regulation through private rules lacks democratic legitimacy. Scientific organizations as private legal entities do not have the authority to issue laws in principle. As registered associations of civil law they can, however, issue statutes according to Section 25 of the German Civil Code, which are binding for their

---

[72] See the guidelines https://www.mpg.de/232129/researchFreedomRisks.pdf. Another code from a scientific organization in Germany is the one of the German Research Foundation ("Freedom of research and responsibility in research": http://www.dfg.de/download/pdf/dfg_im_profil/reden_stellungnahmen/2014/dfg-leopoldina_forschungsrisiken_de_en.pdf).

[73] EURAT project (Fn. 68), p.14, para. III.

members. According to this they normally have organized decision-making, administrative and often compliance and enforcement structures.[74] The legitimacy of a code like the EURAT Code, produced by a group of individual private actors without legal personality, is only rooted in the outstanding expertise of its members in the field of interest and in the need for precise guidance in a special field. The need for guidance is foremost based on the individual assessment of the actors in the group. It should be pointed out, however, that most of the scientific members of the EURAT group are directors of clinical and research institutes and thus bear *inter alia* organizational responsibilities and vicarious liability in their respective departments or institutes.

The Code itself declares the principles and guidelines laid out in it as *de facto* binding for all non-physician scientists who are involved in whole genome sequencing, especially of patient genomes. It points out that the individual researcher within the established sphere of the freedom of research is solely responsible for following the regulations in the Code.

However, the EURAT Code contains no measures of enforcement so that compliance is basically left to the conscience of each and every scientist. Nevertheless, there are guidelines for the implementation of the Code, which can enhance its practical binding effect. According to guideline No. 8 of the Code, the directors of a given research institute must integrate it into the institution's regulatory framework. They must also ensure through routine courses of instruction and training that researchers align their scientific practice with the principles and guidelines of this Code. Furthermore, they must work toward ensuring that the use of shared data and tissue samples complies with measures and guidelines that correspond to those that are formulated in the Code. The steering committees of the respective institutions are additionally responsible for ensuring that the Code is routinely reviewed in order to improve it and to keep it continuously up-to-date with advances in basic research and bioinformatics as well as ethical and legal developments.

---

[74] For the question of whether codes of conduct of scientific organizations are to be considered as acts of administrative law, see the German Ethics Council, 2013, 132 et seq.

Legal Relevance

Private regulations normally rely on involving experts, and so the scientific knowledge can be directly translated into normatively orientating knowledge. One advantage of this is that it can avoid overregulation by the legislator in the field. Nevertheless, although the EURAT Code is legally not binding, it might indirectly attain legal relevance. Legal effects can be experienced regarding liability in civil law and liability based on labour law measures. Regarding evidentiary privilege in criminal law the Code does not develop a legal effect. Last but not least, the Code of conduct can attain legal relevance by implementation through public legal entities such as universities.

The Code aims to reinforce researchers' rights and responsibilities, giving greater capacity for facilitating self-regulation and with more precise formulation in the context of whole genome sequencing of patient genomes, in order to protect the patient as well as the researcher.[75]

Civil liability due to negligence can arise if misconduct in the exercise of reasonable care is established in accordance with Sec. 276 (2) of the German Civil Code.[76] Some scholars argue that guidelines for medical and research practice can neither cause liability nor relieve the persons concerned from their liabilities.[77] However, this is the minority opinion. According to the prevailing opinion the notion of "reasonable care" includes *inter alia* the following of guidelines, which embody standards in a special field of research such as WGS.[78] According to this, personal negligence can be based for instance on the failure to observe risk minimization measures laid down in a code of conduct, although this must be decided on a case-by-case basis.

Whether a specific code of conduct establishes research standards also has to be proven as the case arises. The question of whether the

---

[75] Ibid, p. 14 et seq., para. III.

[76] "A person acts negligently if he fails to exercise reasonable care." Civil Code in the version promulgated on 2 January 2002 (Federal Law Gazette [*Bundesgesetzblatt*] I page 42, 2909; 2003 I page 738), last amended by Article 4 para. 5 of the Act of 1 October 2013 (Federal Law Gazette I page 3719).

[77] Damm, 2009, 13 et seq.

[78] Ibid, 14 et seq. The following of standards can again be described as acting in a way that is in line with medical-scientific knowledge and the experience of professional practice. Hart, 2005.

EURAT Code at this stage of implementation already qualifies as a standard of medical research would need further examination and is beyond the scope of this chapter. There are several arguments, however, which underline its standard-setting potential, such as being the first code in a comparable field, its establishment through outstanding experts in the fields concerned, and its construction based on reference to already established standards of medicine and scientific research. Its reception by other scientific organizations, which exercise their activities throughout the whole country, might indicate its upcoming establishment as a standard in medical research.[79]

Most of the scientific researchers who were involved in the establishment of the EURAT Code wanted it to protect them from having to disclose patients' data to governmental organizations or juridical instances in criminal procedures. Such a protection was not possible through the EURAT Code since there are thorough regulations in German Criminal Law according to *evidentiary privilege*. The purpose of the right to refuse testimony on professional grounds is to protect the relationship of trust between the practitioners of certain professions and those who claim their assistance and expertise in such a relationship. In accordance with Section 53 (1) No. 3 of the German Code of Criminal Procedure, physicians *inter alia* have the right to refuse to give any information pertaining to knowledge concerning the information that was entrusted to them or became known to them in this capacity, both in court and towards other public authorities. This evidentiary privilege is expanded in Section 53a of the Code to include persons assisting and persons involved in the professional activities of those entitled to refuse testimony on professional grounds.[80] According to the prevailing opinion any analogous extension of evidentiary privilege to other occupation groups is inadmissible; it must therefore be determined on a case-by-

[79] Leopoldina. Nationale Akademie der Wissenschaften, 2014, 12 and 19.

[80] Code of Criminal Procedure in the version published on 7 April 1987 (Federal Law Gazette [*Bundesgesetzblatt*] Part I p. 1074, 1319), as most recently amended by Article 5 subsection (4) of the Act of 10 October 2013 (Federal Law Gazette Part I p. 3799).

case basis whether scientific researchers can be legally regarded as belonging to the assisting staff of the responsible physician.

First, the classification of personnel as assisting staff depends on the actual involvement in relevant professional activities. Although the non-physician scientist conducts sequencing of patient genomes on behalf of the responsible physician, the researcher's activities are under no circumstances subordinated, especially when researchers identify and report additional findings that are beyond the task that was ordered. Second, the determining factor for qualifying as a professional assistant entitled to evidentiary privilege is whether one's activities include participation in the relationship of trust between the physician and the patient. This is definitely not the case for non-physician scientists conducting genome sequencing since they never have direct contact with the patient.[81] Thus, Sections 53 and 53a of the Code of Criminal Procedure do not guarantee protection for researchers against the coercive efforts to force testimony as laid down in Section 70 of the Code. In order to secure protection for researchers EURAT suggested considering new legislation to expand the circle of actors who are entitled to evidentiary privilege.[82]

Labour law measurements can embody a solution for the directors of research institutes to integrate the EURAT Code into the regulatory framework of their institutes, as they are required to do according to the EURAT Code. First, the employee can give appropriate instructions to integrate a special code of conduct exercising instruction rights according to Section 106 of the German Industrial Code and Section 315 of the German Civil Code. Though the employer can only specify existing contractual obligations when exercising these rights it is still possible to integrate a code of conduct attached to existing contractual compliance regulations.[83] The second option for integrating a code of conduct into the regulatory employment framework is through contractual agreements. In the case of new employment this solution is less problematic, but much more so where it involves supplementary agreements to already

---

[81] For the conditions for qualifying assisting personnel see Gercke B. § 53 and § 53a. In: Julius/Gercke/Lemke et al, 2009.

[82] See also the Explanations of the EURAT Code (Fn. 68), 28 et seq., para. 6.

[83] HWK/Lembke, §106 GewO Rn. 44. MünchArbR/Blomeyer, §48 para. 32.

existing employment contracts.[84] Standardized individual agreements need to be consented by both parties, which might lead to discrepancies within bigger research institutes. At the same time, integrating ethical and conduct guidelines into the regulatory employment framework might exercise a moral push on researchers to justify their science towards patients as well as to the public in general. Also, indicating the legal relevance of compliance with codes of conduct for civil liability might contribute to an increased willingness on the part of employees to locate their implementation within the contractual regulatory framework. The third option for integration is an agreement between the works council and the directors. Such an agreement is often an indispensable measure for the integration since codes of conducts as compliance regulations are often subject to co-determination.[85]

Once employees are obligated to comply with a code of conduct, non-compliance can lead to a material breach of duties,[86] resulting in labour law-related disciplinary actions and sanctions up to and including dismissal.

Codes of conduct can also gain legal relevance by *measures of implementation* through public entities. Private norms usually become relevant *in* law when the legislator particularly or through general clauses requires compliance with the private regulation.[87] Such an example can be found in Section 137 (f) of the German Social Code, where the legislator explicitly and directly incorporates medical standards for a Disease Management Programme by referring to "evidence-based guidelines"[88] Such a route for the EURAT Code to attain legal relevance is not yet expected, especially since German Law contains no specific regulation on genetic research, nor are there any initiatives to expand the scope of the existing act on medical

[84] Mengel, 2009; para. 31 et seq.

[85] Ibid, para. 50.

[86] Mengel/Hagemeister, 2007.

[87] Damm, (Fn. 80) 12 et seq.

[88] The fifth book of the German Social Code – legal health insurance coverage – (Section 1 of the Social Code of 20 December 1988, BGBl. I p. 2477, 2482), which was last changed through the implementation of Section 5 of the Social Code of 2 December 2014.

genetic diagnosis to translational medicine, where such a reference would be appropriate.[89]

However, Section 8 (5) of the State University Law of Baden-Wuerttemberg grants universities, which are public legal entities, the right to adopt legally binding decisions in order to regulate their matters as far as the legislator has not laid down any rules in respect of these matters.[90] According to this possibility and to Section 19 (1) No. 11 of the State University Law of Baden-Wuerttemberg, which empowers the senates of the universities to make principal decisions on research, the Senate of the University of Heidelberg has decided to adopt the EURAT Code of Conduct and other results published by the group.[91]

Further Advantageous and Disadvantageous Features

In addition to its legal relevance, private self-regulation can positively contribute to scientific development, in ways that outweigh potential disadvantageous consequences. Such regulation can ensure a timely follow-up of specific scientific developments, since issuing private regulations normally takes less time than a legislator would need to intervene in such matters. It is usually performed by experts, which ensures its disciplinary appropriateness, notwithstanding the need to avoid unreasonable scientification. Moreover, private regulations in form of codes of conduct regularly specify ethical standards for the activities concerned and reaffirm these through different legal systems, even if they only concern a limited circle of target groups and only sector-specific issues. Last but not least, the actors involved usually consider codes of conduct to be the most effective way to regulate specific issues.[92]

---

[89] The German Genetic Diagnostics Act of 31 July 2009 (BGBl. pp. 2529, 3672), which was last changed through the implementation of Sections 2 (31) and 4 (18) of the Act of 7 August 2013 (BGBl. I p. 3154).

[90] The State University Law of Baden-Wuerttemberg (Landeshochschulgesetz-LHG) of 1 January 2005, Gbl. p. 1.

[91] See the decision of the Senate of the University of Heidelberg of 28 January 2014.

[92] Sieber/Engelhart, 2014; 116 et seq.

Conclusion

The process of whole genome sequencing crucially requires the active participation of non-physician scientists and other personnel in order to proceed from the initial tissue samples to sequenced DNA and to the evaluation of those sequences. To be able to participate these researchers need benchmarks for conducting their activities.

The EURAT Code of Conduct was developed as a private form of self-regulation to establish and specify the ethical and conduct benchmarks for scientific researchers involved in the sequencing of patients' genomes which previously had been lacking. Being non-law but specifying ethical and practical corridors of decisions and actions when conducting sequencing, in an area in which the legislator has not yet became active, and thus regarding these instances as not conflicting with existing legislation but as going further than the prevailing law was designed to cover, the code contributes to the establishment of a public-private co-regulation, a pluralistic norm system for dealing with WGS in translational research.

The code as a result of private self-regulation can achieve important legal relevance in civil and employment law and also through implementation measures carried out by public entities. Taking these and other advantageous features into account, the code is clearly an important milestone for reflecting scientific development.

Literature

American Academy of Pediatrics Committee on Genetics. Newborn screening fact sheets 83 (1989), 449-454.

American College of Medical Genetics and Genomics. ACMG Updates Recommendation on "Opt Out" for Genome Sequencing Return of Results, 2014. (URL: https://www.acmg.net/docs/Release_ACMGUpdatesRecommendations_final.pdf.)

Borrell, B. US clinics quietly embrace whole-genome sequencing. Nature News, 2010. doi: 10.1038/news.2010.465.

Brand, A., Brand, H. Schulte in den Baumen T. The impact of genetics and genomics on public health, European Journal of Human Genetics 16 (2008), 5-13.

Callaway, E. Norway to bring cancer-gene tests to the clinic. Nature. 12 Feb 2012. doi:10.1038/nature.2012.9949.

Charchenko, T. Phenylketonuria patients are in need of timely treatment, Ukrainian Medical Journal online publication, 5 October 2012. (URL: http://www.umj.com.ua/wp-content/uploads/2012/10/FK.pdf). Accessed 28 January 2015.

Clayton, E.W. Ethical, legal, and social implications of genomic medicine, New England Journal of Medicine 349 (2003), 562-569.

Collins, F.S., McKusick, V.A. Implications of the Human Genome Project for medical science, Journal of American Medical Association 285 (2001), 540–544.

Council of Europe. Convention for the Protection of Human Rights and Dignity of the Human Being with regard to the Application of Biology and Medicine: Convention on Human Rights and Biomedicine, 1999. (URL: http://conventions.coe.int/Treaty/en/Summaries/Html/164.htm).

Cressey, D. UK pushes whole genome sequencing into clinical practice, Nature News Blog. 10 December 2012.

Damm, R. Wie wirkt "Nichtrecht?", Zeitschrift für Rechtssoziologie 30(1) (2009), 3-22.

Siegmund-Schultze, N. 3 Fragen an Prof. Dr. med. Christof von Kalle vom Nationalen Zentrum für Tumorerkrankungen in Heidelberg, Deutsches Ärzteblatt. 17 Feb 2012, 109:A321.

Dryzek, J.S., Holmes, L.T. Post-Communist Democratization: Political Discourses Across Thirteen Countries, Cambridge University Press, Cambridge, 2002.

Epps, Tracey. Regulation of Health Care Professionals. In: Downie, J, Caulfield, T., Flood, C. (eds.). Canadian Health Law and Policy, 3rd ed., Butterworths, Toronto, 2007.

European Commission. Proposal for a REGULATION OF THE EUROPEAN PARLIAMENT AND OF THE COUNCIL on the protection of individuals with regard to the processing of personal data and on the free movement of such data. 2012 (General Data Protection Regulation). (URL: http://ec.europa.eu/justice/data-protection/document/review2012/com_2012_11_en.pdf). Accessed 25 January 2012.

German Ethics Council. Human Biobanks for Research. Opinion. 15 June 2010.

—. The Future of Genetic Diagnosis – from Research to Clinical Practice. Opinion. 30 April 2013.

German Research Foundation. Freedom of research and responsibility in research (URL: http://www.dfg.de/download/pdf/dfg_im_profil/reden_stellungnahmen/2014/dfg-leopoldina_forschungsrisiken_de_en.pdf).

Green, R.C., Berg, J.S., Wayne, W.G. et al. ACMG recommendations for reporting of incidental findings in clinical exome and genome sequencing. Genetics in Medicine 15(7) (2013), 565-574.

Green, Michael, Botkin, Jeffrey. Genetic Exceptionalism in Medicine: Clarifying the Difference between Genetic and Nongenetic Tests, Annals of Internal Medicine 138 (2003), 571-75.

Guttmacher, A.E., Collins, F.S. Genomic medicine: A primer, New England Journal of Medicine 347 (2002), 1512–1520.

Hart, D. Allgemeiner Teil. In: Hart, D. (ed.). Ärztliche Leitlinien im Medizin- und Gesundheitsrecht. Recht und Empirie professioneller Normbildung, Nomos, Baden-Baden, 2005, 23-84.

Hayden, E.C. Is the $1,000 genome for real? Nature 507(7492) (2014), 294-295.

Health Canada. Regulatory transparency and openness. (URL: http://www.hc-sc.gc.ca/home-accueil/rto-tor/index-eng.php). Accessed 30 January 2015.

Heinig, C, Kalle, C. Translationale Onkologie - Effiziente Umsetzung onkologischer Grundlagenforschung, Zentralblatt für Chirurgie 137(1) (2012), 1-3.

Henssler, M., Willemsen, H. J., Kalb, H. J. (eds.) Arbeitsrecht Kommentar, Dr. Otto Schmidt, Köln, 6. new ed., 2014.

International Bioethics Committee. Report of the IBC on the Principle of Non-discrimination and Non-stigmatization. SHS/EGC/IBC-20/13/2 REV.3. 6 March 2014.

Julius, K.P., Gercke, B., Lemke, M. et al. (eds). Strafprozessordnung. 4th edition, C.F. Müller, Heidelberg, 2009.

Khoury, M.J., McCabe, L.L., McCabe, E.R. Population screening in the age of genomic medicine, New England Journal of Medicine 348 (2003), 50–58.

Khoury, M.J. Genetics and genomics in practice: the continuum from genetic disease to genetic information in health and disease, Genetics Medicine 5 (2003), 261-268.

Kohlmann, A. Integration of Next-Generation Sequencing into Clinical Practice: Are We There Yet? Seminars in Oncology 39(1) (2012), 26-36.

Laurie, G., Mallia, P., Frenkel, D.A. et al. Managing Access to Biobanks: How can we reconcile individual privacy and public interests in genetic research? Medical Law International 18(10) (2010), 315-337.

Leopoldina. Nationale Akademie der Wissenschaften. Individualisierte Medizin, Voraussetzungen und Konsequenzen. National Recommendations. December 2014.

Max Planck Society. Guidelines and Rules on a Responsible Approach to Freedom of Research and Research Risks (URL: https://www.mpg.de/232129/researchFreedomRisks.pdf).

Mengel, A., Hagemeister, V. Compliance und arbeitsrechtliche Implementierung im Unternehmen, Betriebs-Berater 62 (2007), 1386-1393.

Mengel, A. Compliance und Arbeitsrecht. Implementierung, Durchsetzung, Organisation, Beck, München, 2009.

Ministry of Health and Social Development of the Russian Federation. The Statement on Newborn Genetic Screening, Ordinance No. 185. 22 March 2006.

Ministry of Health of the Republic of Belarus. Clinical protocols on identification and treatment of phenylketonuria and hypothyroidism, Ordinance No. 781. 7 August 2009.

Molnár-Gábor, F. Integrating Next-Generation Sequencing into Medical Diagnostics - A Snapshot of Normative Challenges. Medicine and Law. World Association for Medical Law 33(4) (2014), 115-126.

Molnár-Gábor, F., Weiland, J. Die Totalsequenzierung des menschlichen Genoms als medizinischer Eingriff – Bewertung und Konsequenzen, Zeitschrift für medizinische Ethik 60(2) (2014), 31-42.

Murken, J., Grimm, T., Holinski-Feder, E. Humangenetik, Thieme, Stuttgart, 2006.

National Human Genome Research Institute. Regulation of Genetic Tests. (URL: http://www.genome.gov/10002335). Accessed 30 January 2015.

Newborn Screening Task Force. Serving the family from birth to the medical home: newborn screening: a blueprint for the future – a call for a national agenda on state newborn screening programs, Pediatrics 106 (2000), 389-422.

Nuffield Council on Bioethics. Genetic screening: a supplement to the 1993 report by the Nuffield Council on Bioethics, Nuffield Council on Bioethics, London, 2006.

Ojha, R.P., Thertulien, R. Health care policy issues as a result of the genetic revolution: Implications for public health, American Journal of Public Health 95 (2005), 385-388.

Peltonen, L., McKusick, V.A. Genomics and medicine: dissecting human disease in the postgenomic era, Science 291 (2001), 1224–1229.

PHG Foundation. Next steps in the sequence. The implications of whole genome sequencing for health in the UK, October 2011 (URL. http://www.phg foundation.org/file/10363/9).

Presidential Commission for the Study of Bioethical Issues. Privacy and Progress in Whole Genome Sequencing. October 2012.

Project EURAT "Ethical and Legal Aspects of Whole Human Genome Sequencing", Cornerstones for an Ethically and Legally Informed Practice of Whole Genome Sequencing: Code of Conduct and Patient Consent Models. Position Paper. Heidelberg, 2013.

Rechel, B., McKee, M. Health reform in Central and Eastern Europe and the former Soviet Union, Lancet 374 (2009), 1186-1195.

Rehmann-Sutter, C. Das Ganze Genom. In: Bartram, C.R., Bobbert, M., Dölling, D. et al. (eds.). Der (un)durchsichtige Mensch, Winter, Heidelberg, 2012, 255-278.

Richardi, R., Wlotzke, O., Wißmann, H., Oetker, H. (eds.) Münchener Handbuch zum Arbeitsrecht, Beck, München, Vol. 1, 2.ed. 2000.

Rosenberg, Roger. Translating Biomedical Research to the Bedside. A National Crisis and a call for action, Journal of the American Medical Association 289(10) (2003), 1305-1306.

Sandroff, R. Direct-to-Consumer Genetic Tests and the Right to Know, Hastings Center Report 40(5) (2010), 24-25.

Scherrer, J. Das Gendiagnostikgesetz, Lit, Berlin/Münster, 2011.

Sieber, U., Engelhart, M. Compliance Programs for the Prevention of Economic Crimes. An Empirical Survey of German Companies, Duncker & Humblot, Berlin, 2014.

Sirpa, Soini. Genetic testing legislation in Western Europe – a fluctuating regulatory target, Journal of Community Genetics 3(2) (2012), 143-53.

Sung, N.S., Crowley jr, W.F., Genel, M. et al. Central challenges facing the national clinical research enterprise, Journal of the American Medical Association 289(10) (2003), 1278-87.

UNESCO. International Declaration on Human Genetic Data. 2003. (URL: http://www.unesco.org/new/en/social-and-human-sciences/themes/bioethics/human-genetic-data/).

UNESCO. Universal Declaration of the Human Genome and Human Rights. 1997. (URL: http://unesdoc.unesco.org/images/0012/001229/122990eo.pdf).

United States, Genetic Information Nondiscrimination Act of 2008, Pub.L, 110–233, 122 Stat. 881.

Venter, J.C., Adams, M.D., Meyers, E.W. et al. The sequence of the human genome, Science 291 (2001), 1304-1351.

Wilson, J.M.G., Jungner, G. Principles and practice of screening for disease, WHO, Geneva, 1968.

Wollenschläger, F. Der Drittbezug prädiktiver Gendiagnostik im Spannungsfeld der Grundrechte auf Wissen, Nichtwissen und Geheimhaltung: Krankheitsveranlagungen im Familienverbund und das neue Gendiagnostikgesetz. Archiv des öffentlichen Rechts 138(2) (2013), 161-203.

Yoon, P.W. Public health impact of genetic tests at the end of the 20th century, Genetics Medicine 3 (2001), 405–410.

7

Genetic Transparency – Transparency of Communication

Gabrielle M. Christenhusz, Lorraine Cowley, Tim Ohnhäuser, Vasilija Rolfes

Introduction

This chapter will focus on ethical issues concerning the communication of genetic transparency. Elsewhere in this book there is a chapter devoted to defining what might be meant by genetic transparency,[1] but in this section we use the term broadly to indicate what can be known about a person's genes, by whom and for what purpose, in three very specific contexts: non-invasive prenatal testing (NIPT), pursuit of genetic diagnosis by parents of children with suspected/potential genetic conditions, and genetic testing for a known cancer predisposition.

In each context we will discuss not only what can be known within the technical limits of new and existing technologies (the extent to which our genetic transparency can be known) but also what meanings of genetic transparency can be or might be attributed to, derived from or assimilated into our cultural, social and working practices.

The first two sections of this chapter focus on communication between a medical professional (in our specific contexts, a gynaecologist and a clinical geneticist) and the parent/s (specifically, the pregnant woman or the parents of children with suspected/potential genetic conditions). The final section of the chapter focuses on communication between family members who have made a choice for or against predictive cancer testing. In all

---

[1] See ch. 1.

cases, we understand communication to be multi-directional, flowing back and forth between the various communicating partners. This communication may encompass both technical aspects (e.g. the genetic test's analytical and clinical validity and clinical utility)[2] and questions of meaning. It is important to acknowledge that various factors may hinder or challenge the communication process. From the patient's point of view (and here we also include parents and family members), these factors will include literacy, personal and family history of illness and loss, recent traumas, and risk tolerance including the availability of social and clinical support.[3] We also note that the various reactions of the family to personal and family experiences of disease and loss are more important than the mere fact of having these experiences; for example, childhood epilepsy may be a shameful trauma in some families, and just a fact of life in others. This leads to a sort of "phenotype-meaning gap", analogous to the genotype-phenotype gap (one genotype can lead to diverse phenotypes, and diverse genotypes can lead to a single phenotype): here, particular phenotypes will be interpreted differently by different people.[4] Thus we acknowledge that those who are communicating with each other may find themselves talking about different things while expecting to be understood. It is from this briefly stated understanding of some of the issues surrounding communication that we approach our case studies in this chapter.

The cases we use to illustrate our discussions are from the contexts of Germany, Belgium and the United Kingdom. The way in which genetic counselling is offered is different in each of these countries; in Germany and Belgium, for example, genetic counselling can be offered by non-specialists such as gynaecologists or other physicians and the specific profession of "genetic counsellor" does not exist, whereas in the UK it would be offered mostly via the genetics service comprising geneticists and genetic counsellors. Furthermore, the regulation of genetic testing across these countries varies. For example, the current German and Belgian models of service provision are self-regulated, while the UK model is

---

[2] Sanderson/Zimmer et al., 2005.

[3] Rolland 1987; Christensen and Green 2013.

[4] Christenhusz/Devriendt et al., 2015.

government-regulated. The different approaches to regulation will produce inter-country variations in what genetic tests are available and how they can be accessed. Therefore the extent to which physicians and counsellors can offer genetic testing will vary and this will influence how communicating options for genetic intervention is framed in the clinic. The scope of this chapter does not permit further elaboration here about the differing regulatory practices, although chapter 6 has set out some of the similarities and differences of self-regulation and government-regulation.

While there are differences in the approach to communication of genetic transparency across these locations there are also core themes that resonate to a greater or lesser extent in each. Working from core themes this chapter seeks to problematise some of the ethics of communicating potential and actual genetic transparency. Some of the thematic questions that have guided the production of this chapter are:

> How do new genetic technologies challenge conventional expectations of communication between physicians and patients?
>
> In the context of NIPT and work with children and parents, who is the expert who should deal with any challenges arising from these situations?
>
> Is having genetic transparency a choice or a responsibility?
>
> What are the potential and actual consequences of communicating genetic transparency?

Through these cases we will explore the problems we have encountered for the notion of genetic transparency in these areas in particular, but other areas will be considered to a lesser extent.

The chapter is structured so that it first explores some of the problems associated with non-invasive prenatal testing in Germany. Next, the experiences of parents seeking a genetic diagnosis for their children in Belgium will be discussed. The final case will highlight some of the social consequences of genetic transparency from a research study conducted in the UK. Our concluding section will highlight some additional questions for ethical consideration.

Non-invasive prenatal testing (NIPT) – Questions that arise in genetic screening of the foetus[5]

The field of prenatal diagnosis is undergoing a technological revolution, which brings challenges for all parties involved – pregnant women and their partners, physicians and other medical experts and, of course, the foetus. A simple blood test without any (physical) risk at an early stage of gestation allows a huge variety of predictions to be made about the foetus, from determining sex and chromosomal status to, presumably, whole genome sequencing. How is this communicated by physicians to pregnant women and what are the challenges? We will argue that due to the relative simplicity of the test, genetic transparency will become apparent in hitherto unsolicited ways. We will also argue that transparency may arise as an unexpected or incidental side effect of our care for the health of the foetus.

This section begins with a short overview of the new technology and its possibilities and then addresses the interaction between physicians and pregnant women, focusing on Germany. We then explore existing structural conditions, including the questions of who is the expert, and whether prenatal testing is viewed as a choice or an obligation.

The new method of non-invasive prenatal testing (NIPT)

In 1997 Dennis Lo and colleagues confirmed evidence of cell-free foetal DNA (cffDNA) in maternal blood, raising the possibility of using the mother's blood to screen foetal DNA as a non-invasive prenatal test (NIPT).[6] On the basis of this DNA located in the maternal plasma, it is now possible to detect numerical chromosomal aberrations such as Down syndrome, trisomy 13 and trisomy 18, monosomy x and other sex chromosome aneuploidies, by analysing a blood sample of the pregnant woman. Using NIPT, which entered the market less than five years ago, it is also possible to identify specific single gene disorders such as cystic fibrosis. In 2012 the sequencing of the whole foetal genome was successfully

---

[5] This part was contributed by Vasilija Rolfes and Tim Ohnhaeuser.
[6] Lo, 1997.

accomplished for the first time.[7] The current tests have a high sensitivity for detecting trisomy 21: about 98.9–100% with a false positive rate below 1%.[8] NIPT is already available for trisomies, monosomy x and foetal sex detection in many countries as an individual health care service and there is strong competition between the few providers that offer the test worldwide. Costs are dropping continuously and it is assumed that the test will become a standard procedure in prenatal care. There are several advantages of the new non-invasive test compared with conventional invasive prenatal diagnostic methods such as amniocentesis. First and foremost, NIPT has no procedure-related health risks for the foetus or the pregnant woman, e.g. there is no additional risk of miscarriage after performing the test. It undoubtedly has the potential to reduce the number of women who undergo invasive testing, because only pregnant women with a positive result after NIPT will have to do so to secure the results.[9] A temporal aspect may also play a positive role: NIPT can be performed as early as the 10th week of pregnancy, which is earlier than conventional screening methods. This has the advantage of giving the woman the opportunity to terminate the pregnancy after a positive result at an early stage of gestation, which can be physically and psychologically less burdensome for her.[10] With the number of potential findings that are being rapidly brought into the process by biotechnology companies, the variety of medical implications increases – only one fact that will challenge prenatal care in the future.

A view of the prenatal counselling setting in Germany

As medical and health care systems differ from country to country, we can also indicate some relevant differences in the way pregnant women are handled in prenatal care, diagnostics and counselling. Aside from these differences, common main principles of medical practice (at least in Western countries) such as informed consent and

---

[7] Kitzman et al., 2012; Lo, 2012.

[8] Palomaki et al. 2011; Bianchi et al., 2012; Norton et al., 2012; Zimmerman et al., 2012.

[9] Benn/Cuckle/Pergament, 2013.

[10] Deans/Newson, 2013; Heinrichs et al., 2012.

shared decision making enable us to discuss the central issues of NIPT in a global sense. Nevertheless, these principles must be also analysed in the context of local conditions. In Germany, the health care setting shows characteristics that offer both risk and opportunities for the interaction between physicians and pregnant women. It is anticipated that NIPT will soon be paid for by health insurance funds, thereby establishing this test as part of everyday practice, and this will not only rapidly increase the sheer number of tests but will also mean that all pregnant women are routinely confronted with the offer of NIPT.

From a structural point of view, gynaecologists play a central part in supporting women medically during pregnancy, accompanying women from the first weeks of gestation until they give birth, usually in hospital. Sometimes other gynaecologists – for example, prenatal experts specializing in ultrasound – are consulted, if further medical examination is needed or desired. Genetic counsellors or genetic nurses in Germany do not have their own job profile, which makes Germany an exception in comparison to almost all Western countries.[11] Human geneticists are trained physicians as well but there are too few of them to provide nationwide coverage, as would be possible with qualified counsellors and nurses. These structural characteristics further strengthen the physician's role in this process. Today, NIPT in Germany is offered by gynaecologists, and the question that arises is whether communication between pregnant women and their gynaecologists will be adequate to deal with the current challenges that accompany NIPT – and the challenges to come, since the development of NIPT trends towards more extensive genetic screening.

So, for pregnant women in Germany there is usually one medical expert: her gynaecologist. Of course, this exclusive relationship has many advantages, such as there being a single contact person for all medical questions, which avoids possible conflicts, compared to a situation in which counselling and information from several medical experts differ and could lead to making women feel insecure. On the other hand, this close interdependence also bears risks which need to be pointed out, especially in the light of the new technologies. Of

---

[11] Schmidtke/Rueping, 2013.

course, informed consent and shared decision making should be guaranteed, but instead of describing the whole process we are focusing on the area of "information" in this context. It is obligatory for physicians to inform women about all further diagnostic possibilities, once pregnancy has been determined. So the way in which information is transmitted as well as the time and quality of information are of vital importance for the pregnant woman's decision making about further prenatal diagnosis. Emerging themes from our empirical work suggest that, despite regulation of the provision of information, physicians vary widely in *how* they inform women, *when* and about *what*. This would probably not be the best precondition for the implementation of an emerging technology like NIPT.[12] One central question about the new technology: does every gynaecologist have enough professional expertise, e.g. in molecular genetics and genetic diseases, to inform women sufficiently about non-invasive blood tests, the results and potential consequences? In short: who is the expert? And who should be allowed to deal with NIPT? The risk of a procedure that is fast, risk-free, easily available and – in the near future – inexpensive is that it will become very widespread without the consequences being thought through and without having enough experts available. The question is not fundamentally about communicating information; it is more about communicating the potential *consequences* of information. Against the background of the lack of genetic counsellors mentioned above, the question of whether every gynaecologist has enough expertise to meet the requirements becomes more acute.

Ensuring high-quality counselling will present a practical and technical challenge in the years to come. Women's expertise has not yet been considered, and the hierarchy within physician-patient interactions is one reason for that.[13] But the situation in Germany is at least one step behind other Western countries, one reason being that the first question to be answered is: who is actually the expert in the medical field?

---

[12] Results from our project "Indication or Information? The physician's role in the context of non-invasive prenatal diagnosis" (funded by the German Federal Ministry of Education and Research) have not yet been published.

[13] BZgA, 2006.

In a setting in which the woman is not well informed, she could begin by expecting "health information" about her baby, and could be confronted with the option of termination as the only "therapy" – another strong argument for establishing extensive counselling by experts *prior* to the blood test.

Challenges for communication in the context of NIPT and prenatal diagnosis in general

Prenatal diagnosis presents two main difficulties to communication. First, there should be adequate and sufficient education and counselling of pregnant women before and after prenatal diagnosis. This might not sound innovative today from developed countries' points of view, but in Germany at least there is reasonable doubt about whether these basic requirements are always fulfilled. The (sparse) empirical data in this field show, for example, that women often do not know afterwards that they have undergone prenatal diagnosis. Most of the pregnant women think the different examinations are part of routine prenatal care, which is a false assumption (only three ultrasound examinations are included in standard care); clearly, this is not informed choice. If they are undergoing prenatal diagnosis without knowing it, we have to ask how their real informed choice can be guaranteed. Most pregnant women believe that the examinations help to promote the health of the foetus, which is their strongest argument for wanting different kinds of diagnostic tests. They are not aware that further prenatal diagnostics also serve to detect genetic variations for several diseases.[14] Put bluntly, with increasing genetic prenatal diagnosis the issue will be not how to look after someone (the foetus), but will be about pregnancy prevention and abortion.

The second point concerns what can be and what should be communicated about the foetal DNA constitution in prenatal diagnosis – and who should be allowed to inform women about it. There are some concerns expressed in the scientific literature on prenatal non-invasive screening in relation to communication between the physicians and pregnant women . Keeping all the new

---

[14] Ibid.

opportunities of NIPT findings in mind, how can the woman's informed choice be secured without overtaxing her with too much detailed pre-test information? This again bears the risk of an "information overload", which could threaten the woman's ability to make autonomous reproductive choice.[15] On the other hand, there are concerns that physicians and other health care workers could share the opinion that pregnant women need to be less informed about the non-invasive method of giving "just" a blood sample, compared to more invasive methods like amniocentesis.[16]

The advantages of NIPT may also cause pressure on the physicians: "Given the reduced risk of foetal miscarriage associated with NIPT, physicians may feel obligated to offer prospective parents NIPT for a wide range of genetic disorders and conditions not currently part of the prenatal screening protocol to avoid a potential suit for wrongful birth."[17] From this point of view physicians are willing to communicate information about the foetal DNA that is medically ambiguous, or even to serve non-medical purposes, in order to avoid being accused of withholding information. This undermines the idea of shared decision making and together with other risks could lead to a "forced transparency": not originally intended by anyone, and mainly at the expense of the weakest party in this process – the foetus. The extent to which the foetus can be seen as an "agent" during pregnancy is a complex question that will not be discussed here. But the issue of whether the foetus can be seen as a patient has a more specific relevance in this context. In our expert interviews we asked, "who is the patient (in prenatal care)?" and received a wide range of responses. Without going into too much detail, it is clear we need to know who our patient is, if we are dealing with tests that may potentially lead to genetic transparency.

Public attitudes always carry the risk of putting pressure on the pregnant woman's decision making.[18] We assume, keeping in mind discussions about disability, that a new offer of prenatal diagnosis will increase public pressure to make use of it. For pregnant women,

---

[15] de Jong et al., 2011.
[16] van den Heuvel et al., 2010; Deans/Newson, 2011.
[17] King, 2011, 621.
[18] Benn/Chapman, 2010.

declining a (physically) risk-free blood test is more difficult than declining amniocentesis. Shakespeare (1998) discussed the way in which ready availability of prenatal testing created medical and cultural expectations to have babies without any impairment, and how declining such tests could be judged as "selfish".[19] At the same time, the decision to bear a disabled child will probably not become easier. This is one aspect where the concept of (reproductive) choice is influenced by an assumed obligation for testing often driven by vague public opinions.

Moral pressure on the pregnant woman also arises from the terminology used. "Vorsorge" (prevention) during pregnancy implies a kind of moral imperative. It is not easy to decline prevention in health issues, especially for pregnant women who bear the responsibility for an unborn child. If we connect this expected pressure with the empirical results that women mostly see any kind of prenatal diagnosis as standard or routine prevention, shortcomings in the concept of autonomous choice become obvious again. Furthermore, the meaning of "genetic prevention" then needs to be discussed.

Physicians who work in the field of prenatal diagnostics are particularly involved in these reflections. They would be responsible for obtaining the genetic information about the foetus and would then have to decide how to present it to the pregnant woman. Since the clinical pathway of diagnosis, prognosis and therapy does not fit into the prenatal genetic testing context, they have to be aware that, from the woman's perspective, increasing uncertainty due to increasing genetic risk could lead to terminating her pregnancy as a preventive measure.

How could these undesirable developments – "information overload" or "forced transparency" – be avoided? Are there any arguments for limiting information? And how could we define a normative framework for counselling in the prenatal context? Clarifying the physician's role and discussing a modification of non-directiveness is one approach.[20] Stringent specifications for and limits to the implementation of NIPT are necessary. In the Netherlands and

---

[19] Shakespeare, 1998, 667.

[20] Rehmann-Sutter, 2009.

the UK, for example, national boards for issues concerning NIPT have been established, ensuring an open regulatory process by monitoring the new technology scientifically.[21]

Non-invasive prenatal testing – Impacts on genetic transparency

In this section we present facts about the new technology of NIPT and give some suggestions for the debate that needs to take place on genetic testing of the foetus during pregnancy. In addition to the ethical and legal questions that have been discussed generally in the preceding chapters, specific NIPT-related questions arise. Genetic testing in a non-prenatal context usually takes place after extensive counselling that tries to ensure autonomous decision making. After weighing up the pros and cons it is (ideally) a conscious step to become tested – and transparent – in this specific way. In the NIPT context, as mentioned above, there is a much higher risk of being tested "incidentally", for many reasons. NIPT may become an "offer you can't refuse"[22] – heavily advertised by powerful providers, which all women, most of them unprepared, will be presented when they become pregnant. The genetic testing process is camouflaged here by promoting the "health" of the child, and the most significant difference from other tests is its targeting of the foetus. This is again linked to the central discussions in this book, for example about the right to know and not to know. Is it the parents' right to let the foetus be tested? What questions will arise in the coming years, when more and more children are born being genetically sequenced? At the moment it is mostly "only" a targeted screen of chromosomal alternations, but other findings (beyond single gene disorders or micro-deletions) will almost certainly follow, and it is possible to imagine full genetic transparency of the foetus in only a few years. In the communication between the parties involved, will a genetically transparent foetus be connected with "health"? For these unresolved questions, a lack of regulation of communication, the negation of the foetus' rights, and also responsibility being exerted as a moral pressure, we can speak of a "forced genetic transparency" of the

---

[21] For more information see the official websites: http://www.niptconsortium.nl; www.rapid.nhs.uk.

[22] Schmitz/Netzer/Henn, 2009.

future child. This has the potential to affect interpersonal relations in a myriad of unforeseen ways.

Genetic transparency in the context of parents requesting clinical genetic testing for their children[23]

Many genetic disorders first manifest in childhood.[24] This means that many of those who visit genetics clinics are children and their parents (or legal guardians). The parents of children with a genetic disorder are generally seeking two types of genetic information: the causal genetic variation of their child's condition, with the hope that this will lead to a diagnosis and a prognosis; and reproductive information on the probability that they will conceive other children with the same condition. They may also receive other kinds of genetic information (variously called incidental findings, unexpected results, unsolicited findings, and secondary variants) about their children, themselves, or other family members.

The aim in this section is to explore communication and genetic transparency in the context of parents requesting clinical genetic testing for their children. It is assumed that these are children who already show symptoms of a genetic disorder. Predictive genetic testing of children for adult onset conditions remains a separate, contentious issue.[25] We will start by taking each of the three types of genetic information that can be discovered as a result of a clinical genetic test (causal variants, reproductive information, and other information), and discuss what genetics professionals and parents expect to be able to communicate in each case. These expectations of communication will be compared with the current state of knowledge, to assess how realistic they are. In a second step, the expected and unexpected consequences of genetic transparency will be examined.

---

[23] This part was contributed by Gabrielle M. Christenhusz.
[24] Biesecker/Green, 2014.
[25] Clayton/McCullough et al., 2014.

It should be noted that the genetics professionals mainly referred to in this section are clinical geneticists, not genetic counsellors or any other type of genetics professional. This is because the qualitative research drawn on in this section was conducted in Belgium with parents whose children had undergone genetic testing[26], and in that country clinical geneticists are parents' first point of contact in the world of genetics. The qualitative research consisted of semi-structured interviews with parents about their experiences of genetic testing and their hopes and expectations for the return of results, both expected (causal variants and reproductive information) and unexpected (other information).

Communication surrounding the causal variant

The main reason that parents visit a genetics clinic and genetic professionals conduct genetic tests is to identify the causal variant of the child's condition. This can be referred to as the "primary finding" of a genetic investigation, to distinguish it from all other possible findings that are not the primary focus of the test.[27] When parents arrive at a genetics clinic with their children, the parents and the genetics professionals are expecting to communicate with each other about the causal variant for which they will test. However, their expectations of how the causal variant, once identified, will be able to be used, will differ somewhat.

From the point of view of genetics professionals, they expect that conducting a genetic test will enable them to communicate the identity of the causal genetic variant and thereby diagnose the child's genetic disorder. A diagnosis is necessary for optimal clinical management, both now and as the condition progresses. Genetics professionals are thus also hoping that by identifying the causal variant, they will be able to give information about the child's prognosis. From the point of view of parents, they will expect a genetic test to enable them to communicate about diagnosis, prognosis and optimal clinical management for their child. It became apparent in interviews that many parents also hope that identification

[26] Christenhusz/Devriendt et al., 2014.
[27] Christenhusz/Devriendt et al., 2013.

of a causal variant will enable them to put a name to their child's condition.[28] While genetics professionals focus on the genetic identity of the causal variant, parents focus on the identity and characteristics of the genetic condition. However, both hope to be able to communicate about the child's condition. For genetics professionals, this may be limited to the clinical aspects of the child's condition. Parents may take a wider view and want to understand the non-clinical effects of the child's condition too. Genetics professionals will be the "experts" on the clinical uses of identifying the causal variant, but it is parents who will be the experts on the effect that identifying the causal variant will have on their child's life and the life of the family. While most of our interview parents expressed admiration that their clinical geneticists treated them and their children as "whole people," listening to the parents' expertise regarding the child[29], some clinicians can find it a challenge to be open to an exchange of expertise.

How realistic are the hopes of genetics professionals and parents about the primary finding? The most recently reported success rate of clinical sequencing for the identification of a causal variant is approximately 25%[30], though this is improving with increased clinical use of new genomic techniques. This means that the "primary finding" is not "found" in most cases. In cases when it is found, there often remains a gap between identification of a causal variant and being able to use this information. This is because of two factors shared by many of the childhood onset genetic disorders that require genetic testing: a high level of locus heterogeneity, meaning multiple causal genes, which can interact with each other in a range of ways; and/or a high level of clinical heterogeneity, meaning the children show a broad range of symptoms and presentations.[31] Identifying a causal variant does not always coincide with identifying the genetic condition. This does dampen the hopes and expectations of genetics professionals and parents somewhat, despite the existence of a growing number of success stories about how the identification of

---

[28] Christenhusz/Devriendt et al., 2014.

[29] Ibid.

[30] Yang/Muzny et al., 2013.

[31] Kingsmore/Saunders, 2011; Schuettpelz et al., 2011.

the causal variant led to improved clinical management, sometimes dramatically so. This calls into question what genetics professionals and parents will be able to communicate about the causal variant and its relationship to the child's genetic condition.

Genetics professionals and parents may not dwell on the challenges involved in identifying the causal variant, with everyone hoping that they will be amongst the "25%" of cases in which a result is found. However, the difficulties involved in the genetic testing of rare childhood conditions do seem to be acknowledged in consultations. For example, many interview parents mentioned the number of doctors and specialists they had first had to see with their child before ultimately being referred to a genetics clinic.[32] These parents expressed understanding for other parents who give up before reaching the stage of a genetic test, because the search for a genetic diagnosis is so difficult for rare childhood genetic conditions. While the interview parents felt a certain responsibility on behalf of their child to find a genetic diagnosis, no one generalised that responsibility to all parents of children with genetic conditions, nor did they judge other parents for not pursuing testing. Genetic testing at least in their context was seen as a genuine choice.

Communication surrounding reproductive information

While the primary topic of communication will be the causal variant for the child's condition, this may lead to communication about the reproductive relevance of the causal variant. A secondary reason for visiting a genetics clinic will often be to discover the recurrence risk, for the parents or other family members, of having another child with the same condition. Genetics professionals can estimate the recurrence risk based on the family history and the severity and extent of the condition. The ultimate hope expressed by many clinicians and parents, gleaned from interviews, is that a causal variant will be found that can then be screened for prenatally, thus ensuring that no other child be born in that family with the same genetic

[32] Christenhusz/Devriendt et al., 2014.

condition.[33] In other words, the ultimate hope is to be able to communicate reproductive information with certainty. Until a causal variant is found, clinicians tend to advise that parents delay having other children, and parents can tend to accept this advice. (In our admittedly small sample of 17 families, for instance, only one of the affected children was the oldest child in the family; in all other families, either the affected child was the youngest child, or the affected child was the only child and parents were waiting on a causal variant in order to continue having unaffected children.) That is, knowledge of the recurrence risk is often not adequate reproductive information for clinical geneticists and parents, because any risk is considered to be too high.

The difficulties in identifying the causal mutation have been described above. There is also an additional hurdle: whether the healthcare system will allow the prenatal screening of individual mutations, from an ethical, justice or financial point of view. The desires of genetics professionals and parents for reproductive information that can be used to ensure healthy children are thus challenged by technical and practical factors. Furthermore, the possession of reproductive information may not be enough to ensure healthy children, as discussed in the previous section on prenatal testing. It is unclear how these challenges to the limitations and meaning of reproductive information are communicated between genetics professionals and parents. Some of the interview parents laughingly said they had been "forbidden" by clinical geneticists to try to have more children until the causal variant had been identified in their severely disabled child.[34] While this was said in a half joking way, it was almost certain that none of the parents had talked with their geneticist about the possibility that no causal variant would be found. The closest I came in interviews was one mother who broke down in tears at the thought that she might keep waiting so long for the assurance that she could have a second child who would be healthy that it would end up being too late to have another child, and if her disabled daughter then died from the condition (the geneticist had been unable to give a prognosis), she would be left with no children.

---

[33] Ibid.

[34] Ibid.

It was obvious that this mother had never been given the opportunity to express these fears to her geneticist.

Communication surrounding other information

A recent document published by the (US) Presidential Commission for the Study of Bioethical Issues divides the "other information" that may become available after conducting a genetic test into various categories: anticipatable and unanticipatable incidental findings, secondary findings, and discovery findings.[35] All of this "other information" can be characterised in the first place as falling outside the primary aim of the investigation (denoted the "causal variant" or "primary finding" above), and in the second place as having potential relevance for the patient. "Incidental findings" (IFs) is the more traditional term, though there is growing dissatisfaction with it.[36] Anticipatable IFs, as the name suggests, can be anticipated by the nature of the test. For example, genetic testing involving a child and their parents can be expected to identify misattributed paternity if it exists. Genetic testing that broadly involves "looking for causal mutations" can be expected to identify common causal mutations familiar to all geneticists, such as *BRCA-1* and *BRCA-2* mutations associated with increased risk of breast and ovarian cancer. Secondary findings, also called secondary variants, are mutations known to be associated with treatable or preventable genetic disorders. Though unrelated to the primary clinical question, they are actively sought for each time a genetic test is conducted. The American College of Medical Genetics and Genomics (ACMG) led the way in April 2013 by publishing a list of secondary variants that they recommended be actively sought for each time one of the new genomic tests (whole genome or exome sequencing) is conducted in a clinical context.[37] Other groups are trying to compile their own lists. The idea behind a list of secondary variants is not "opportunistic screening," as some have termed it[38], but more the attempt to take full advantage of new

---

[35] Presidential Commission for the Study of Bioethical Issues, 2013.
[36] Christenhusz/Devriendt et al., 2013.
[37] Green/Berg et al., 2013.
[38] Middleton/Patch et al., 2014.

technology. The final category in the Presidential Commission's list, discovery findings, is more relevant to research than clinical contexts.

The amount that genetics professionals will be able to communicate to parents about unanticipatable IFs and discovery findings will be limited to the simple fact that these two types of "other information" may be obtained when conducting genetics tests. Genetics professionals will be able to be more explicit in their communication about anticipatable IFs and secondary findings. The only danger is that of "information overload": of having so many possible anticipatable IFs and secondary findings to communicate that parents become overwhelmed.[39] In contrast, parents are less likely to raise the topic of "other information" explicitly in their communication with genetics professionals. Some may be so focussed on their child's current genetic condition that the disclosure of anything else comes as a shock; for these parents, the possibility of other information will have to be presented to them. Others may have assumed that the aim of genetic testing is to identify any and all mutations related to genetic disorders, not just the causal mutation; for these parents, the specific aim of the genetic test will have to be distinguished from the possible but by no means certain discovery of other information.

It is therefore the genetics professional who will have to take the lead in communications about the other information that may become available from genetics tests. At the same time, the genetics professional should be open to listening to the parents speaking from their position of expertise on the effect that receiving this other information may have on themselves, their child and the wider family.[40] Genetics professionals may limit other information to that which is clinically actionable; parents, on the other hand, may view other information as anything that they can "use" in a broader sense.[41] This might include information that can help with practical and emotional preparations before the predicted genetic condition manifests.

---

[39] Ali-Khan/Daar et al., 2009.

[40] Christenhusz/Devriendt et al., 2015.

[41] Christenhusz/Devriendt et al., 2014.

Linking this discussion back to the previous section on prenatal testing, NIPT is not aiming at any specific causal variant, because there is no specific genetic condition to be diagnosed. Reproductive information as discussed above does not really fit the aims of NIPT either; the focus is not future possible pregnancies, but the present pregnancy. The genetic information aimed at in NIPT is information about the health of the foetus. As explained above, "health" is a very broad concept. In a way, everything and nothing that can be discovered through NIPT falls under "other information," as it is unclear what is and is not to be expected.

Expected and unexpected consequences of genetic transparency

Genetics professionals and parents expect that genetic testing of the child will result in transparency about the causal variant; either the identity of the causal variant, or the assurance that the child does not have a genetic condition. This last expectation is almost impossible to fulfil, as the lack of a causal variant at the present time does not rule out a genetic condition. Challenges regarding the identity of the causal variant have been detailed above.

Moreover, as mentioned above, parents expect or hope that genetic transparency about the causal variant will enable them to put a name to their child's condition. Interview research suggests that parents expect that if they can do this, several benefits will follow.[42] Having such a name would provide a sort of reassurance to parents that they are not "inventing" their child's problems. More importantly, parents can use the name when dealing with the "outside world," be it family, friends, strangers, or the healthcare and education systems. Having a name is perceived as being useful in explaining the child to other people, and in arming the parents against the criticism that their child's condition is somehow the fault of bad parenting techniques. A name is also often required in order to access certain necessary benefits offered by the healthcare and education systems. Sometimes parents found that just being able to name the mutation ("my child has duplication X," for example) was enough.

---

[42] Ibid.

The ultimate hope remains the name of a syndrome, with clearly recognisable features.

Another consequence of identifying the causal variant may not be expected, or even desired: if it is shown that the causal variant associated with the child's genetic disorder has been inherited from one or both parents. Even if the mutation is recessive in the parents, the possession of this mutation can raise issues of guilt and self-identity for the carrier parent. If the mutation is not recessive but has variable penetrance and expression, questions of identity and of "what is normal?" can be intensified for the affected parent. Interview parents responded in various ways to the news that their child shared a mutation with one of the parents.[43] A few saw it as an explanation for their own difficulties in childhood, despite the best efforts of the clinical geneticist to explain that there may be no connection.

Some genetics professionals and parents will expect that genetic testing of the child will result in transparency regarding reproductive information. This transparency can apply to the child, their parents, and other family members. If the causal variant associated with the child's genetic disorder is shown to be a spontaneous mutation, this provides reproductive information relevant to future children in the family – the child's own children (in cases where the affected child is able to have children of their own), any future children of their parents, and potentially the future children of their siblings. Parents involved in our interviews recognised the complexity of reproductive information for multiple family members, including the challenges of whether and when to tell children that they are (potential) carriers and what to do if not everyone in the family wants to know.[44] Many interview parents reported wanting to know the causal variant of their child's condition so they could use it themselves for reproductive purposes and also pass on this information to all of their children, so that all relevant parties would have the same reproductive options; these parents were also very open about how sensitive and difficult such discussions might be for their children. An unexpected or at least potentially undiscussed consequence of genetic transparency

---

[43] Ibid.

[44] Ibid.

about reproductive information is that genetic information could come to light about the grandparents, and possibly about other older members of the extended family (aunts, uncles etc.).

Genetic transparency regarding the "other information" that might be discovered in the course of genetic testing is not the aim. When reflecting on potential consequences of genetic transparency regarding this last type of genetic information, interview parents suggested that both positive and negative consequences might be possible.[45] Potential positive consequences include the ability to take certain preventive measures if the information is disclosed in time, with an improvement in the child's subsequent health. Potential negative consequences include treating their child in a way they would rather not if they knew a particular piece of genetic information (e.g. becoming overly protective, or being unable to let the child just enjoy their childhood), and how the child's ability to act or make plans might be restricted if a particular piece of genetic information were known.

Is it the genetic transparency of the *child* or of the *future* that is the parents' main focus? In many cases, the distinction will not matter, as it is the future of the child that is the parents' concern. However, it is an interesting distinction when one considers that the genetic transparency of the past and present is often easier to illuminate than that of the future. In most cases, genetic information only speaks about potential futures; genetic predisposition is not predestination. Even a mutation such as that associated with Huntington's disease, which is 100% predictive, cannot pinpoint when the condition will arise or how it will progress. The genetic transparency of the future will always be limited, just as the transparency of the future in general is. However, most communications about genetic information seem to focus on the future, and what will be able to be done with the genetic information to change the present and/or the future.

The genetic transparency of the past and the present is easier to demonstrate than that of the future. It encompasses such issues as whether the child inherited their condition from one or both parents, and whether the parents are in fact the child's biological parents. However, if the emphasis remains on the future, it is possible that

---

[45] Ibid.

these other issues will tend to be brushed over until they become realities – that is, until a father discovers that he is not the child's biological father, or a parent discovers that they are not so "normal" after all. Because our ideas of the past and present are very deep-rooted and usually unquestioned, tied up with our conception of our own identity, any challenges to these ideas through a new-found genetic transparency will be a major shock. More attention needs to be paid to how to prepare genetics professionals and parents for the past and present becoming genetically transparent in unexpected and possibly unwanted ways.

Choice or responsibility? Seeing through genetic transparency[46]

The previous sections have illustrated how the availability of new technologies in genetic diagnosis adds to the understanding and construction of the term "genetic transparency". In the domains of non-invasive prenatal diagnosis and the paediatric genetics service, the communication of whatever new genetic information can be gleaned from new technologies has been problematised from several perspectives. Concerns have been highlighted about limitations to the meaningful interpretation of results, the ethics of disclosing incidental findings, and the rights of those tested. The problems of communication have therefore focused largely on the problems posed for health professionals and how transparent they should be about the limitations and uncertainties of new genetic information. What, however, might it mean to a person within their family to choose or not to choose to have a genetic test, thus becoming (more) genetically transparent? I will explore this within a very specific context.

I will first highlight the research from which I have taken the empirical data; I will then explore how the narrative methods used contributed to the production. These data would not have become transparent without the distinct methods used, and this places narrative as a potentially important vehicle to discussing and communicating genetic transparency. This section will explore from a

[46] This part was contributed by Lorraine Cowley.

social perspective what it might mean to communicate genetic transparency within families.

About the study

The qualitative research that informs this section was undertaken in the North of England. It was inspired by my genetic counselling experience and my interest in families with a cancer predisposition gene causing Lynch Syndrome (LS). People with LS have an increased lifetime risk of developing cancers of the bowel, the digestive and urinary tracts, and in women, cancers of the endometrium and ovaries.[47] Those identified as having a mutation are referred for bowel surveillance colonoscopies every eighteen months to two years from the age of twenty-five years. Since there is no proven effective surveillance for gynaecological cancers, women who have a mutation are additionally invited to discuss cancer preventative surgery (removing their ovaries and uterus) from the age of thirty-five years as a means of managing their increased risk of gynaecological cancers, providing they do not want to have (more) children. In the previous sections, the technologies being discussed are used to screen for unforeseen health issues or to try to define and explain a known health issue; in this situation, the genetic technology is able to pinpoint a specific known family mutation in a mismatch repair gene, and therefore diagnoses the specific condition LS. Although in one sense this may be considered to afford a greater sense of certainty than the screening and exploratory technologies, it is important to emphasise here that, as Christenhusz has stated previously in this chapter, the potential for future genetic transparency in LS is limited because the risks of developing cancer are probabilistic and not certainties.

Other scholars have suggested that the context of genetics can variously affect personal and family lives. Conveying genetic knowledge within families can frame hierarchies of communication.[48] For example, it has been shown to matter within families who tells who about genetic conditions and the onus of communicating

---

[47] Lynch et al., 2005.
[48] Clarke et al., 2001a; Clarke et al., 2001b.

information does not always lie with the person affected by the condition; parents for example, regardless of their genetic status, have been framed as holding a hierarchical position in this respect. Personal lives of individuals may be affected by experiencing genetic guilt irrespective of them having a genetic trait[49]; or the context of testing can trouble the concept of "chosen" kin by focusing on shared, and perceived immutable biology[50]. Placing families in contexts where they make decisions to have a genetic test for known diseases can frame senses of family obligation that are important in maintaining and creating a moral identity.[51] It is against this emotive background of intimate personal relationships, shared understandings of duty and responsibility that kinship is framed and against which I will explore the significance of communicating genetic transparency within families.

The aim of the research was to explore meanings and senses of family generated at the intersections of genetics and kinship. It focused on a family known to a Regional Genetics Service since the 1970s, who contributed to research[52] characterising one of the genes (hMLH1) causing LS. This genetic research made testing for LS possible. It framed individuals from the biologically constructed family as the first to know their genetic risk status for this condition, and the first to have genetic transparency in this aspect. My study emerged out of an interest in what this involvement in genetic research and knowledge production might have had on participants' sense of individual and kinship identity. In-depth multiple narrative interviews, using visual props such as family photographs, their genetic family tree and social maps, were conducted with fifteen of fifty members of the family who had been offered testing (some of whom had the genetic mutation, while the majority did not). Respondents were invited to discuss their family relationships and their engagement with genetic testing. Findings showed how participants discursively managed both biomedical notions of family as given and contemporary notions of family as chosen, when

---

[49] Murakami et al., 2001; Marteau, 1999.
[50] Finkler et al., 2003; Strathern, 2003; Finkler, 2000.
[51] Denier, 2010; Chipman, 2006; d'Agincourt-Canning, 2006; Hamilton, 2004.
[52] Kolodner, Hall et al., 1995.

drawing their complex boundaries of family. What is pertinent to the notion of genetic transparency is how senses of morality were framed and intensely highlighted in participants' narratives about those who declined a genetic test. It is at the intersection of choice and responsibility that a moral lens appeared through which participants viewed those who declined testing, and it is within this emotive setting that I will consider problems of communicating genetic transparency.

Making genetic transparency a moral issue

Only when exploring the possibility of choice in genetic testing did senses of morality become transparent in the narratives of the participants. Whilst all participants felt that choice was important, it became apparent that there were common moral imperatives with respect to children, to self-care and to research, suggesting that participants shared a moral framework for genetic testing. To illustrate these points I will share some participants' narratives. Although these have been selected for their clarity of expression, all participants expressed these themes, regardless of gender, age and whether they had the mutation or not.

Graham was a man in his sixties who did not have the mutation. He discussed genetic testing in the following way:

> Graham: Well I think if you've got children, right, they need to know what the situation is. So to some extent that is a responsibility. I think if you haven't got children, it's up to you, But in my case, even if I hadn't got children, I'd have wanted to know. I think you're better off. My general view is anyway you're better off knowing than not knowing. Irrespective if a thing is bad, you're better off knowing and that's always been my view.

While this quote illustrates the moral imperative to be tested for the benefit of the children, it also demonstrates a moral imperative just "to know." From his perspective, if a test exists, knowledge is available and therefore there is a moral pressure "to know."

Frank was a man in his forties who also tested negative for the mutation, and he framed his decision as a choice made after logical and rational consideration.

> Me: Okay, can you remember what influenced you to have the test?

> Frank: Just, well my sort of background that I've got with (work) and sort of data; I'm dealing in facts and data all of the time. So I'd want to know what the outcome of that would be. And then, from then, sort of make or assess what the options were, if you like. You know? I've got a choice then of do I, do I want to go for the test. What if I got the results of the test, do I then want to go for check ups and so on and so forth? And just being armed with the knowledge I suppose that if anything did happen, you've given yourself the best chance of being able to fight it I suppose.

Frank frames knowledge as power that one can "arm" oneself with to "fight" adversity, which here is the possibility of developing cancer. This has parallels with work done on certainty, uncertainty and risk.[53] In trying to give himself the best "chance"[54], he is possibly trying to claim a sense of control by attempting to minimise risk. By appealing to logic, he is framing a genetic test as the logical thing to do and therefore his "choice" to be tested creates a self-identity as a rational human being and responsible valued citizen.[55] The rational human being, as depicted in this data, is one who does everything in their power to avoid disease.[56] The importance of looking after oneself by minimising the risk of becoming ill is set against a background of first and second-hand experiences of losing family members to inherited genetic cancer. The responsible citizen depicted in this data is one who looks after their family and fulfils family obligations to care for children and all those they call close kin. These obligations of responsible citizenship can only be fulfilled if an individual has primarily cared for themselves by maintaining their health, and in this data this is what links the arguments of rationality to responsible citizenship.

Ian, a man in his sixties who tested negative, also appeals to logic in reflecting on his decision to be tested but draws on the argument that genetic testing is for the "greater good", thus introducing a moral

---

[53] McLaughlin/Goodley, 2008; Lupton, 1999.
[54] Denier, 2010.
[55] Chapple et al., 2008.
[56] Rose, 2007.

imperative not only to be altruistic, but also to be a good biological citizen.[57]

> Me: Okay and do you still feel that you did the right thing in having the test?
>
> Ian: Oh without a shadow of a doubt! You've got to. God I can't understand people (who do not have a test). This is why medically we're doing all what we're doing. It's for the benefit of people. And when you're not going for these tests, to me you're refusing. You're refusing the help. You're refusing help. I can't understand that. You know I can't understand that. But the logic behind that, I can't see it. I can't see the logic in that – No…

Later Ian goes on to say that aside from helping himself, another reason for having a genetic test was to help others:

> Ian: And the other one (reason) was well I can do a bit of good… you know… if that helps other people which it's got to do…

The premise of caring for yourself, to make sure you can fulfil your family and kinship obligations, is implicitly extended to wider society here. A contribution to the development of society is constructed, not only in the context of caring for one's own, but also through the assumption that others will learn from one's medical experience. Furthermore, Ian goes on to give a generational emphasis to his constructed argument that genetic testing is worthwhile. This generational perspective implies that to have knowledge of a disease equates to having influence over the course of a disease and this is synonymous with cancer prevention messages about early detection and cure.[58] Thus in this context it does not matter so much that the future is not as genetically transparent as the past or present, because the chances of improving one's lot are constructed as remaining valid even when cancer is only discussed as a probability rather than a certainty.

> Ian: But I can't see anybody saying it would be a bad thing that like. But you'll maybes know better than me, with

---

[57] Rose, 2007, 39-40.

[58] Mowat/Jupp, 2008; Raffle et al., 2003.

> interviewing more people. But (it is) common sense again for me. Knowledge (audible exhalation), you know (you should) get to know and do something about it! Simple as! Thirty year ago you had no choice, you were stuck with what you had and that was it. Now you've got a choice. You know?

Distinct from but linked to the potential to control his fate, Ian's quote speaks of a normative attitude that people should unquestioningly take advantage of tests developed through medical knowledge. A perception of control is formed, and a choice to have a genetic test becomes a choice to exercise control over one's fate. Previous generations are framed as powerless and therefore non-culpable in accepting their fate whereas, paradoxically, present generations do have a "choice" but by making the "wrong choice", to decline intervention, become culpable for their fate. This underpins the concept that those who "choose" to decline testing are authors of their own misfortune, and these stories were also told in this family. Denier talks about the boundary between chance and choice as being the spine of conventional morality and that any serious shift in this is dislocating.[59] In framing choice to have a genetic test as "common sense", choice is in reality negated and a moral imperative appears whereby those who do not follow "common sense" are judged irrational

What it might mean to remain genetically opaque

The previous section has painted a moral landscape in which decisions or choices to have a genetic test have been sketched. I now discuss in more detail what it meant in the study family to *decline* a genetic test and, by declining, remaining genetically opaque. There were eight known test decliners in the family and participants discussed them and their decisions not to be tested in moral terms. They were negatively judged for not fulfilling kinship responsibilities, being selfish, cowardly, illogical, lacking in character and foolish. The visual motif with which participants depicted those who declined a genetic test was the ostrich. They were those that "buried their heads

[59] Denier, 2010, 101.

in the sand". This was framed as a morally indefensible position and posed problems in communicating genetic transparency within the family. Whilst everyone felt that test decliners had the freedom of choice to say no, there was a sense that they had chosen "wrongly". I illustrate with an example, in which Diane gives an insight into the family dynamics of pressures to have a genetic test; both her mother and her aunt declined to be tested.

Firstly, Diane highlights that to become genetically transparent is a choice:

> Diane: I think it has to be an individual choice, yeah. So I would say it was individually on each of us to decide whether to go or not for it. Yeah, definitely. But we [she and her siblings] did speak about it.

This imperative of individual choice was interesting, given that Diane's parent exercised that individual choice to decline testing. This was problematic for Diane and her siblings in ways that were visible in the next quote.

> Diane: Well, it was my mother and aunt's belief (not to be tested), I suppose and their opinions and all that, but I mean, it didn't really help me.
>
> Me: So how did that make you feel about your (parent) and your Auntie (both declined testing)?
>
> Diane: I think (sigh) I think initially we were all, probably all of us (Diane and her siblings) were a little bit annoyed. But then I think they (her mother and her aunt) were annoyed with us for going to get tested.

Although it may have been "an individual choice" this quote gives a sense of the tensions that existed between those who chose to be tested and those who chose not to be tested. The notion of choice is rarely experienced in isolation but is contingent upon many factors, and it is interesting that the individuality of choice was framed as a fundamental right in this context, where contingencies of choice such as senses of family obligation, sensitivities to others' needs were still perhaps considered prior to genetic testing but were not perceived as being acted upon. Although those who declined genetic testing also declined to participate in this research, other participants had narratives about them. Diane gave a sense of family dynamics being disturbed when individuals exercised their choice to be, or not to be, tested. For Diane, her parent's decision to decline was framed as a lack of help for herself. Her parent's choice should have been morally contingent upon senses of family obligation such that, in having a test, her parent would be putting the needs of her children above her own. Although from Diane's perspective her mother failed to fulfil this obligation, she later discursively managed and framed her mother's decision not to be tested as one that not only had consequences for herself, but also for medical and family research ("now we'll never know"). She represented her parent as being annoyed by Diane's own decision to be tested. I registered Diane's frustration and annoyance that her parent had declined testing.

I think it is significant that those who declined a genetic test, who remained genetically opaque, also declined to participate in this study. I would argue that the strong moral voices articulating the imperative to be genetically transparent are what silence the genetically opaque. To add to the problems of communicating genetic transparency, then, there is an additional problem of communicating that one is *not* genetically transparent, and this could potentially lead to hitherto unconsidered social inequalities in the age of genomic medicine.

What does narrative give to the notion of communicating?

In this research I do not claim that there was one objective reality, narrative or story accounting for the experiences of this family. Instead, I claim that there are multiple realities, stories and narratives that, in their plurality, helped to illuminate some social and personal consequences of genetic testing for cancer predisposition syndromes. In taking one person's view and contrasting it with another, I was not looking for corroboration of fact but for shared representation or variant representation. That was not to say that one account was more valid than the other. Each was interesting in and of itself and had an additional dimension of interest when placed in relief with other accounts. I do not, therefore, claim that the findings of this project are generalisable to other individuals' experiences but are illustrative of the kinds of understandings that can and have been generated in this particular context.

In the frontier breaking world of genomic medicine it is easy to be swept along by the hyperbole of the new, the exciting and the plethora of dominant voices in government and health care settings that are (self) appointed spokespersons for the future wellbeing of humankind. What we need to attend to are the missing voices from those debates and domains. Whose voices are silenced? The methodology of narrative analysis was critical to producing this data. Only in multiple readings of participants' narratives did the missing voices become clearly apparent. From the data generated by this research, the test decliners are the silent voices. A concern therefore for debates about genetic transparency is that, if we decline an opportunity to become genetically transparent, do we then become "morally transparent"?

If narrative is a useful means of communicating and elucidating the notion of genetic transparency, then I would argue that what is also absolutely required is not only a forum for narrators but also a forum for critical listeners. Critical listeners should be given the space to reflect the narratives back to society and highlight the missing voices. The most pressing question for critical listeners would be how to engage the missing voices in public narratives about genetic transparency.

Conclusion

We conclude this chapter with a few reflections and questions that bring the three sections together.

First, we made the important observation that before one can consider the communication of genetic transparency in a specific context, we need to reflect on what genetic transparency means in that case. The three sections of this chapter came out of three distinct genetic contexts: prenatal screening; non-targeted diagnostic testing of children; and targeted diagnostic testing of adults in a family affected by a genetic condition. As we saw, the concept of genetic transparency means different things in different contexts. Extrapolating NIPT to other screening contexts, it can be posited that the idea of genetic transparency will be most vague in genetic screening contexts, in which the transparency aimed at is more about health than about genetics, and with the concept of "health" not always clearly defined (cf. the "health" of the foetus). Genetic transparency remains broad and uncertain in non-targeted genetic testing contexts, in which unexpected transparency may arise, not just for the individual patient but also possibly for their family, in particular their parents. The concept of genetic transparency is most clear-cut in targeted genetic testing, in which genetic transparency equates to the knowledge of a positive or negative result (accepting that the consequences of the result may not be so clear-cut), but then the question of moral transparency arises. What remains the same across all contexts is the existence of layers of uncertainty, especially uncertainty about the future (e.g. the timing and progression of genetic disorders). The two ideas of genetic transparency and uncertainty or its opposite, certainty, came together at various points in the three sections above. In the NIPT section, expectant mothers and gynaecologists were seeking genetic transparency of the foetus in order to be certain about the future health of the foetus. In the section on parents of children with a suspected/potential genetic disorder, parents and genetic professionals were seeking genetic transparency of the child in order to be certain about diagnosis and prognosis (in terms of the causal variant) and reproductive risk. In the section on genetic testing for a known cancer predisposition syndrome, family members were seeking genetic transparency to be certain about their own risk of developing cancer, in doing so caring for their children and altruistically helping society. This raises several questions:

- Are we chasing certainty by looking for genetic transparency? Is this an elusive dream? (Given for instance the impossibility of assuring expectant mothers that their future child will be "completely healthy"; the difficulty of finding a causal variant in the case of rare childhood diseases; and the impossibility of saying with certainty that someone with a genetic predisposition will ultimately develop cancer. All of which pose key challenges for any effectively working communication.)
- To what extent do we rely upon or expect certainty in our medical discussions or communications between genetic professionals (or medical professionals in general) and patients/parents? How do the limitations and uncertainties of genetic transparency challenge this?

Second, established genetic transparency can raise questions of moral transparency, although the latter is often not acknowledged in clinical settings. The third section of the present chapter dealt explicitly with the issue of moral transparency. Turning to the first two sections, it can be postulated that genetic transparency for the children (arising out of prenatal testing or diagnostic testing) can lead to questions of moral transparency for the parents; questions of potential blame if the genetic transparency of the children points towards something inherited from one or both parents, and questions of potential responsible parenthood if parents decide not to allow their children (born or unborn) to become genetically transparent. It is not always clear who decides who will become transparent. Is it doctors advising or putting pressure on parents and patients, or is it a free choice from the parents (but what about their child's choice?) and patients? What impact do legislation, policy and societal norms and expectations have? Again, these questions trouble the notion of effective communication about genetic transparency. An additional question is: what influence do moral judgements have on the decision to become transparent? This question can be considered in these contexts within the concepts of blame or virtue ("responsible parenthood", "responsible citizenship") if one does or does not opt for transparency. It has been argued that once moral pressure comes to bear on a previously "free choice," choice becomes an obligation.[60] With the increasing geneticisation of medicine, genetic testing may become the "norm". It would be interesting to investigate the relationship between the normalisation of genetic testing and moral pressure; at what point will genetic transparency become an offer "too *good* to refuse?" And what would it mean for moral identity if one refused?

Third, while a full consideration of informed consent in the context of new genetic and genomic testing is beyond the scope of this chapter, we finish by raising a few relevant points. Just as some forms of genetic transparency are not foreseeable (e.g. incidental findings), which raises challenges for the traditional informed consent process[61], some of the consequences of genetic transparency are also not foreseeable (e.g. moral judgement). It is difficult to include that which is unforeseeable in an informed consent process in any meaningful way (the only option seems to be blanket consent). To move forward towards a new way of understanding informed consent is to be open to different forms of expertise, beyond the clinical and

[60] Denier, 2010.

[61] Christenhusz/Devriendt et al., 2015.

medical expertise of the genetics professional, and to be open to a more tolerant, flexible and emancipatory communication setting. The unforeseeable consequences of genetic transparency include interpersonal ones – the impact of genetic transparency on family relationships, including but not exclusive to relationships between children and parents. The question of abortion is another consequence of genetic transparency that is rarely discussed explicitly in genetics consultations. It is the patient or parent/s who is the expert in terms of what the interpersonal consequences of genetic transparency will mean for them and their family, not the genetics professional. The question arises of how complementary expertise can become part of a new way of doing informed consent. What is needed is negotiated consent between genetics professionals and patients/parents about the extent of transparency: this refers both to the extent of transparency that is *aimed at* (ranging from a single-gene disorder to the full genome) and the extent of transparency that is *possible* (given technical and practical limitations). Discussion of the extent of transparency should also take into account the consequences of transparency, so that either party has the opportunity to say that they would prefer limiting the extent of transparency before genetic testing is carried out. It is obvious that these fundamental discussions are time-consuming and unlikely to be practicable in a traditional communication setting between patient and (medical) expert. A more universal approach of communicating genetic transparency is required, which satisfies the needs for information, autonomy and expertise of all parties involved.

Literature

Ali-Khan, S., A. Daar, et al. Whole genome scanning: Resolving clinical diagnosis and management amidst complex data, Pediatric Research 66(4) (2009), 357-363.

Benn, P., Howard, C., Pergament, E. Non-invasive prenatal testing for aneuploidy: current status and future prospects. Ultrasound in Obstetrics and Gynecology (2013) 42, 15-33.

Benn, P., Chapman, A. Ethical challenges in providing noninvasive prenatal diagnosis, Current Opinion in Obstetrics and Gynecology (2010) 22, 128-134.

Bianchi, D. et al. Whole genome maternal plasma DNA sequencing detects autosomal and sex chromosome aneuploidies, Prenatal Diagnosis (2012) 32 (Suppl. 1), 3-4.

Biesecker, L.G., Green, R.C. Diagnostic clinical genome and exome sequencing, New England Journal of Medicine 370(25) (2014), 2418-2425.

Bundeszentrale für gesundheitliche Aufklärung (BZgA). Schwangerschaftserleben und Praenataldiagnostik. Repraesentative Befragung Schwangerer zum Thema Praenataldiagnostik, 2006.

Chapple, A., Ziebland, S., Hewitson, P., McPherson, A. What affects the uptake of screening for bowel cancer using a faecal occult blood test (FOBt): A qualitative study, Social Science & Medicine 66(12) (2008), 2425-2435.

Chapple, A., Ziebland, S., McPherson, A. Stigma, shame, and blame experienced by patients with lung cancer: qualitative study, British Medical Journal 328(7454) (2004), 1470-1473.

Chilibeck, G., Lock, M. et al. Postgenomics, uncertain futures and the familiarisation of susceptibility genes. Social Science and Medicine 72 (2011), 1768-1775.

Chipman, P. The moral implications of prenatal genetic testing, Penn Bioethics Journal 2(2) (2006), 13-6.

Christenhusz, G.M., Devriendt, K. et al. Secondary variants: In defense of a more fitting term in the incidental findings debate, European Journal of Human Genetics 21 (2013), 1331-1334.

Christenhusz, G.M., Devriendt, K. et al. The communication of secondary variants: Interviews with parents whose children have undergone array-CGH testing, Clinical Genetics 86 (2014), 207-216.

Christenhusz, G.M., Devriendt, K. et al. Ethical signposts for clinical geneticists in secondary variant and incidental finding disclosure discussions. Medicine, Health Care and Philosophy (2015). Epub ahead of print.

Christensen, K.D., Green, R.C. How could disclosing incidental information from whole-genome sequencing affect patient behavior? Personalized Medicine 10(4) (2013), 377-386.

Clarke, A., Featherstone, K. and Atkinson, P.A. Lay beliefs and the disclosure of personal genetic information to family members, Journal of Medical Genetics 38 (2001a), 42-42.

Clarke, A.J., Featherstone, K. and Atkinson, P. The disclosure of genetic information to family members, European Journal of Human Genetics 9 (Supplement 1) (2001b), C038.

Clayton, E.W., McCullough, L.B. et al. Addressing the ethical challenges in genetic testing and sequencing of children, American Journal of Bioethics 14(3) (2014), 3-9.

D'Agincourt-Canning, L. Genetic testing for hereditary breast and ovarian cancer: Responsibility and choice, Qualitative Health Research 16(1) (2006), 97-118.

Deans, Z., Newson, A. Ethical considerations for choosing between possible models for using NIPT for aneuploidy detection, Journal of Medical Ethics 38 (2012), 614-618.

Deans, Z., Newson, A. Should Non-Invasiveness Change Informed Consent Procedures for Prenatal Diagnosis? Health Care Analysis 19 (2011), 122-132.

de Jong, A. et al. Advances in prenatal screening: the ethical dimension, Nature Reviews Genetic 12 (2011), 657-663.

Denier, Y. From brute luck to option luck? On genetics, justice, and moral responsibility in reproduction, Journal of Medicine and Philosophy, 35(2) (2010), 101-129.

Finkler, K. *Experiencing the New Genetics – Family and kinship on the medical frontier.* Philadelphia, Pennsylvania Press, 2000.

Finkler, K., Skrzynia, C. and Evans, J.P. The new genetics and its consequences for family, kinship, medicine and medical genetics, *Social Science and Medicine* 57 (2003), 403-412.

Green, R.C., Berg, J.S. et al. ACMG recommendations for reporting of incidental findings in clinical exome and genome sequencing. Genetics in Medicine **15**(7) (2013), 565-574.

Hamilton, R. *Experiencing predictive genetic testing in families with Huntington's disease and hereditary breast and ovarian cancer,* Dissertation: Thesis, University of Wisconsin-Madison, 2004.

Heinrichs, B. et al. Ethische und rechtliche Aspekte der Praenataldiagnostik: Herausforderungen angesichts neuer nicht-invasiver Testverfahren, Medizinrecht 30 (2012), 625-630.

King, J. And Genetic Testing for All…The Coming Revolution in Prenatal Genetic Testing, Rutgers Law Journal 42 (2011), 599-658.

Kingsmore, S.F., Saunders, C.J. Deep sequencing of patient genomes for disease diagnosis: when will it become routine? Science Translational Medicine **3**(87) (2011), 87ps23.

Kitzman, J. et al. Non-invasive whole genome sequencing of a human fetus, Science Translational Medicine 4(137) (2012), 137ra76.

Kolodner, R. et al. Structure of the human MLH1 locus and analysis of a large hereditary nonpolyposis colorectal carcinoma kindred for MLH1 mutations. Cancer Research 55(2) (1995), 242-248.

Link, D.C., Schuettpelz, L.G. et al. Identification of a novel TP53 cancer susceptibility mutation through whole-genome sequencing of a patient with therapy-related AML. JAMA 305(15) (2011), 1568-1576.

Lo, Yuk-Ming Dennis et al. Presence of fetal DNA in maternal plasma and serum, The Lancet 350(9076) (1997), 485-487.

Lo, Yuk-Ming Dennis. Non-invasive prenatal diagnosis by massively parallel sequencing of maternal plasma DNA, Open Biology 2(6) (2012). doi: 10.1098/rsob.120086120086.

Lupton, D. *Risk,* London, Routledge, 1999.

Lynch, H., Jass, J., Lynch, J., Attard, T. Hereditary colorectal cancer: An updated review. Part II: The Lynch Syndrome (Hereditary Nonpolyposis Colorectal Cancer), Gastroenterology and Hepatology 1(2) (2005), 117-133.

Marteau, T. Communicating genetic risk information, British Medical Bulletin 55 (1999), 414-428.

McLaughlin, J., Goodley, D. Seeking and rejecting certainty: Exposing the sophisticated lifeworlds of parents of disabled babies, Sociology 42(2) (2008), 317-335.

Middleton, A., Patch, C. et al. Position statement on opportunistic genomic screening from the Association of Genetic Nurses and Counsellors (UK and Ireland). European Journal of Human Genetics (2014).

Mowat, E., Jupp, M. Bowel cancer screening: the story so far, Primary Health Care, 18(3) (2008), 25-29.

Murakami, Y., Gondo, N., Okamura, H., Akechi, T., Uchitomi, Y. Guilt from negative genetic test findings, American Journal of Psychiatry 158 (2001), 1929.

Norton, Mary et al. Non-invasive chromosomal evaluation (NICE) study: results of a multicenter prospective cohort study for detection of fetal trisomy 21 and trisomy 18. American Journal of Obstetrics and Gynecology 207(2) (2012), 137.e1-137.e8

Palomaki, G. et al. DNA sequencing of maternal plasma to detect Down syndrome: an international clinical validation study, Genetics in Medicine 13(11) (2011), 913-920.

Raffle, A., Alden, B., Quinn, M., Babb, P., Brett, M. Outcomes of screening to prevent cancer: analysis of cumulative incidence of cervical abnormality and modelling of cases and deaths prevented, British Medical Journal 326 (2003), 901-904.

Rehmann-Sutter, Christoph. Why Non-Directiveness is Insufficient: Ethics of Genetic Decision Making and a Model of Agency, Medicine Studies 1(2) (2009), 113-129.

Rolland, J. S. Family illness paradigms: Evolution and significance. Family Systems Medicine 5(4) (1987), 482-502.

Rose, N. *The Politics of Life Itself – Biomedicine, Power and Subjectivity in the Twenty-First Century,* Woodstock, Princeton University Press, 2007.

Sanderson, S., Zimmern, R. et al. How can the evaluation of genetic tests be enhanced? Lessons learned from the ACCE framework and evaluating genetic tests in the United Kingdom, Genetics in Medicine 7(7) (2005), 495-500.

Schmidtke, Joerg, Rueping, Uta. Genetische Beratung: Nichtaerztliche Personen koennen ein Gewinn sein, Deutsches Aerzteblatt 110(25) (2013), A1248-A1250.

Schmitz, Dagmar, Netzer, Christian, Henn, Wolfram. An offer you can't refuse? Ethical implications of non-invasive prenatal diagnosis, Nature Reviews Genetics 10 (2009), 515.

Shakespeare, Tom. Choices and Rights: Eugenics, genetics and disability equality, Disability & Society 13(5) (1998), 665-681.

Strathern, M. Re-describing Society, Minerva 41 (2003), 263-276.

van den Heuvel, Ananda et al. Will the introduction of non-invasive prenatal diagnostic testing erode informed choices? An experimental study of health care professionals, Patient Education and Counseling 1(78) (2010), 24-28.

Yang, Y., Muzny, D.M. et al. Clinical whole-exome sequencing for the diagnosis of mendelian disorders. New England Journal of Medicine 369(16) (2013), 1502-1511.

Zimmermann, Bernhard et al. Noninvasive prenatal aneuploidy testing of chromosomes 13, 18, 21, X, and Y, using targeted sequencing of polymorphic loci, Prenatal Diagnosis 32(13) (2012), 1233-1241.

About the Authors

Kirsten Brukamp is Professor of Health Sciences at the DHBW Cooperative State University, Heidenheim, Germany. Current research interests include the history, theory and ethics of medicine and health care, as well as philosophy of the life sciences.

Gabrielle M. Christenhusz completed a PhD in bioethics at the University of Leuven in 2014. The project was a collaboration between the Centre for Biomedical Ethics and Law and the Centre for Human Genetics. Her research interests include the ethics of new genetic technologies, incidental findings, and families.

Lorraine Cowley, a genetic counsellor at the Northern Genetics Service in Newcastle and social researcher, now funded by the Wellcome Trust, is also a visiting scholar with the Policy, Ethics and Life Sciences Research Centre (PEALS) at Newcastle University, UK. Current research interests include the ethics of genetic testing for cancer susceptibility and how moral agendas are framed and responded to in the clinic.

Malte Dreyer is associated with the Institute of Philosophy at the University of Marburg. In 2008-2010 he was visiting lecturer in the Department of Sociology at the University of Kassel and in 2012-2013 Scientific Officer for the project "Genetic Transparency" at the Institute for History of Medicine and Science Studies at the University of Lübeck, Germany. Current research interests include philosophy of science and the relationship between narrative and knowledge.

Jeanette Erdmann is Professor of Cardiovascular Genetics and director of the Institute for Integrative and Experimental Genomics at the University of Lübeck, Germany. Current research interests include the genetics of complex diseases, especially coronary artery disease and myocardial infarction.

Andrei Famenka is a legal medicine specialist in the Legal Medicine State Service of the Republic of Belarus, and teaches health law at Belarusian State Medical University and research ethics in the

Advanced Certificate Program in Research Ethics in Central and Eastern Europe. He is a member of the National Bioethics Committee of the Republic of Belarus. Current research interests include the ethics of international health research, social justice in health care, and the impact of post-Communist transition on the development of research ethics and clinical ethics in the countries of Central and Eastern Europe.

Teresa Finlay is currently a researcher in the sociology of genomics at Cardiff University. She is also an academic visitor at HeLEX, University of Oxford, and a lecturer in research methods for health and social care at Oxford Brookes University. Current research interests include the sociological aspects of direct-to-consumer genetic testing, and sociological influences on emerging health technologies.

Caroline Fündling is a lawyer who works in a law firm in Frankfurt am Main. She is especially interested in employment law and medical law. Current research interests include legal issues of the rights to know and not to know, in particular in the doctor-patient-relationship.

Shannon Gibson is Research Associate at the University of Toronto Faculty of Law. Her research concerns health law and policy, with a particular focus on pharmaceutical policy and the regulation of new health technologies.

Cathy Herbrand is Research Fellow in Sociology at the Reproduction Research Group at De Montfort University (UK) and also a member of the Belgian Advisory Committee on Bioethics. Her research interests lie in the sociological and anthropological study of family, biotechnologies and genetics, with a particular focus on the regulation of new forms of parenthood. Her current research looks at reproductive choices in the context of mitochondrial disorders.

Angeliki Kerasidou is a Researcher in Global Health and Research Ethics at the Ethox Centre, Nuffield Department of Population Health, and the Ethics Coordinator of the Malaria Genomics Epidemiology Network (MalariaGEN). Her academic background is in theology and philosophy and her current research interests include the ethics of genomic research in developing countries, ethical issues of austerity and commercialisation in health care, and research professionalism.

Lene Koch is a historian and Professor of Medical Science and Technology Studies at the University of Copenhagen. Her research interests include past and present uses of human reproductive technologies, genetic technology, the ethics of human genomics, and animal studies.

Fruzsina Molnár-Gábor, Dr. iur., is a lawyer and a senior research fellow at the Heidelberg Academy of Sciences and Humanities and a visiting scholar at the Max Planck Institute for Comparative Public Law and International Law, Heidelberg. Current research interests include the normative analysis of clinical sequencing, data protection, and stem cell research.

Tim Ohnhäuser is research assistant at the Institute for History, Theory and Ethics in Medicine in Aachen (RWTH University), Germany. He is currently working on a project on the physician's role in non-invasive prenatal diagnosis.

Christoph Rehmann-Sutter is Professor of Theory and Ethics in the Biosciences at the University of Lübeck, Germany, and is also a visiting professor at King's College, London. Current research interests include the anthropology of genomics, the ethics of transplantation of blood stem cells between siblings, and moral issues in end-of-life care.

Benedikt Reiz is a PhD student at the Institute for Integrative and Experimental Genomics, University of Lübeck, Germany. His current research focus includes computational biology and the genetic background of complex diseases.

Vasilija Rolfes is a research assistant at the Institute of History, Theory and Ethics in Medicine, RWTH Aachen University, Germany. Her current research interests include the stigmatization of people suffering from psychiatric disorders, and ethical issues in prenatal testing.

Sara Tocchetti received her PhD from the London School of Economics, working on the DIYbio network, socio-technical utopias, and theories of technology-driven social change. She is currently working on the application of "hands-on" practices in science mediation and the radical science movement.

Index

Printed in the United States
By Bookmasters